NEXT WAR

Anything is better than indecision. We must decide. If I am wrong, we shall soon find out and can do the other thing. But not to decide wastes both time and money and may ruin everything.[1]

General of the Armies of the United States Ulysses S. Grant

NEXT WAR

Reimagining How We Fight

JOHN ANTAL

CASEMATE
Pennsylvania & Yorkshire

First published in the United States of America and Great Britain in 2023. Reprinted in 2024 by
CASEMATE PUBLISHERS
1950 Lawrence Road, Havertown, PA 19083, USA
and
47 Church Street, Barnsley, S70 2AS, UK

Copyright © 2023 John Antal

Paperback Edition: ISBN 978-1-63624-335-1
Digital Edition: ISBN 978-1-63624-336-8

A CIP record for this book is available from the British Library

All rights reserved. No part of this book may be reproduced or transmitted in any form or by any means, electronic or mechanical including photocopying, recording or by any information storage and retrieval system, without permission from the publisher in writing.

Printed and bound in the United Kingdom by CPI Group (UK) Ltd, Croydon, CR0 4YY
Typeset in India by DiTech Publishing Services

For a complete list of Casemate titles, please contact:

CASEMATE PUBLISHERS (US)
Telephone (610) 853-9131
Fax (610) 853-9146
Email: casemate@casematepublishers.com
www.casematepublishers.com

CASEMATE PUBLISHERS (UK)
Telephone (0)1226 734350
Email: casemate@casemateuk.com
www.casemateuk.com

Contents

Foreword by Senator Mike Rounds, Member of the US Senate Armed Services Committee		vii
Preface: Who Thinks, Wins!		ix
Acknowledgements		xiii
1	A Failure of Imagination	1
2	The Transparent Battlespace	13
3	The First Strike Advantage	31
4	Top Attack	47
5	Artificial Intelligence and the Accelerating Tempo of War	55
6	The Transition to Fully Autonomous Weapons	67
7	The Kill Web	75
8	The Super Swarm	85
9	Visualize the Battlespace	95
10	Decision Dominance	111
11	The First Starlink War	125
12	Preparing for the Next City Fight	135
13	The Big Blue Blanket—Light Tactical Aircraft for Counter Unmanned Aerial Systems Combat	153
14	Developing a Hybrid Human–Robotic Force	161
15	Command Post Rules	169
16	Forging Battleshock	181
Endnotes		201
Glossary		219
Recommended Resources		225
Selected Bibliography		227
Index		235

US soldiers deploy a small unmanned aerial system at the National Training Center at Fort Irwin, California, in 2022. Behind them is a robotic squad vehicle. The US Army is slowly creating a hybrid human–robotic force to win the next war. (US Army photo)

Foreword by Senator Mike Rounds, Member of the US Senate Armed Services Committee

War is organized violence to gain power and influence. It is ageless, complex and ever-changing. John Antal has written a must-read examination of the changing methods of the next war that anyone involved in the defense of the United States of America should read.

As a member of the US Senate Armed Services Committee and Ranking Republican Member of its Cybersecurity Subcommittee, a large portion of my duties concerns national defense and emerging defense technology. In the past two years, John Antal has briefed senior military leaders, staffs, and me on the changing methods of warfare. His briefing to me showcased the Second Nagorno-Karabakh War with its profound implications for the future of warfare and emerging military technologies. The information he presented showed what we must do to keep ahead of our adversaries. He has conducted these briefings as a patriot, without charge, because he believes in our nation and the cause of freedom.

A lifelong student of warfare, John Antal is well-suited to bring this information to senior national defense leadership and the public. He served 30 years as an Army officer and is a graduate of the United States Military Academy at West Point, the Army Command and General Staff College, and the Army War College. He served as Special Assistant to the Chairman of the Joint Chiefs of Staff, and commanded combat units from platoon to brigade level. After his retirement from the Army, he led technology teams in the commercial sector and later served as a member of the Army Science Board from 2018–21. He is a best-selling author and has published 17 books on military history, leadership and military affairs. Most importantly, he engages leaders to ask the hard questions, and he challenges his audience and readers to think.

The United States is in a dangerous chapter of our national security history. Russia, the People's Republic of China (PRC), Iran, and North Korea threaten peace, the United States, and our allies. The most destructive war in Europe since World War II is being fought following the 2022 Russian invasion of Ukraine, and the PRC, Iran, and North Korea are coalescing in their support of Putin's act of aggression. The Ukrainians are fighting heroically, but the conflict continues and Putin's regime seems determined to conquer their free and independent nation. The PRC continues to threaten its neighbors over territorial boundaries and has sworn to reunite the Republic of China (Taiwan) by **any** means. President Xi has told his military to be prepared for war with the conquest of Taiwan in mind. While supporting Putin's war on Ukraine, Iranian leaders have vowed the destruction of Israel. North Korea remains a provocative and unpredictable nuclear power. Kim Jung Un, North Korea's leader for life, continues to expand his arsenal of nuclear weapons and ballistic-missile programs. The possibility of American forces finding themselves in combat against a peer power, by design or mistake, appears to be growing every day. Weakness, not strength, inspires these regimes to commit provocative acts and aggression at the expense of the United States, our allies, and partners. To keep the peace, we must prepare for war while hoping it never comes.

Foresight in military affairs is rare. As John has clearly said, gaining foresight requires thinking, questioning, and dialogue. *Next War* will help stimulate the thinking, questioning, and dialogue we need to revitalize our efforts to defend the United States. If we commit our forces to war, we do not want it to be a fair fight. Our forces must have superior personnel, technology, and weapons systems to those of the enemy as well as a doctrine that enables those personnel to use that technology and those systems to achieve victory on the battlefield. That requires asking the tough questions raised by John Antal today. If *Next War: Reimagining How We Fight* stimulates a dialogue on the key issues facing our national defense, then it will have served a noble purpose. We must reimagine how we fight, and *Next War* is a good place to start.

Senator Mike Rounds (R–SD) was elected to the United States Senate in 2014 after serving two terms as governor of South Dakota. During his full time in the Senate, he has been a member of the Senate Armed Services Committee (SASC). He was the first chairman of SASC's Subcommittee on Cybersecurity and since 2021 has served as the subcommittee's ranking Republican member. With his strong interest in emerging technologies, he is co-chair of the United States Senate Artificial Intelligence Caucus.

Preface: Who Thinks, Wins!

> … most people would die sooner than think—in fact, they do so.[1]
> BERTRAND RUSSELL

Who thinks, wins.[2] This has always been the case in war. In Homer's *Iliad,* when the warriors of Greece boarded a vast fleet of ships to wage war against mighty Troy, their plan was to meet the enemy and defeat the Trojans in open battle. Without thinking it through, the Greeks committed their armies, fortunes, and honor to a war that continued for 10 long years. The war became a stalemate with neither side able to gain a decisive advantage. After a decade of blood and indecision, and with no end in sight, one Greek leader, Odysseus, described by the goddess Athena as a "great tactician," proposed a clever plan. Odysseus realized the Greeks would never breach the sturdy walls of Troy by brute force alone. They needed a new and imaginative way to win the war. In Virgil's epic poem, the *Aeneid,* Odysseus's plan is described as the scheme of "that inventive head"[3] and a cunning ruse. Odysseus was not only a skillful leader and the Greeks' best strategist, he was also blessed by the gods with foresight. Foresight is the ability to see and fix problems in the short term and create solutions for the long run. Odysseus overcame the walls of Troy using the subterfuge of the Trojan Horse, bringing victory to the Greeks. The lesson from this story is clear: he who thinks in imaginative and surprising ways, wins.

Today, we live in a time of exponential change. Technology is transforming everything, especially the methods of war. Most Americans are unaware of the rapid changes in warfare. With a major war raging in Europe, and with the potential of that war to escalate into a greater conflict, we must pay attention. Communist China appears ready to attack Taiwan. North Korea is saber-rattling. Iran is on the march, fighting a hybrid war with proxy forces and threatening to attack the US and destroy Israel. In late June 2022, the British Army's Chief of the General Staff, General Sir Patrick Sanders, called this time "our 1937 moment,"[4] drawing a comparison to the years leading up to the gathering storm of World War II. It seems the winds of war are howling again. Are we ready for the coming storm? Will we use the time we have wisely before the gale hammers us?

Military and national security leaders have a duty to understand and study the changing methods of war. We count on these leaders to deter war and, if deterrence fails, to fight smart and win. In 1973, Israeli leaders believed the next war would look like the previous war. That war, the "Six-Day War" which occurred from June 5 to 10, 1967, was a decisive Israeli victory that demonstrated the superiority of the Israel Defense Forces. Military superiority, however, is never eternal. From October 6 to 25, 1973, Israel faced a new war that was as much a surprise as it was a severe test. Known as the Yom Kippur War, as it started on the Jewish holy day of Yom Kippur; Israel eventually prevailed, but it was initially a close-run affair. As the Egyptian, Syrian, and other Arab forces mostly used Soviet equipment and tactics, military leaders in the US saw the Yom Kippur War as a corollary for war against the Soviets in Europe and carefully analyzed the conflict. American leaders like Gen. William DePuy, Gen. Donn Starry and others studied the war and learned. They applied these lessons to the US Army and this foresight changed how the Army trained, equipped, and fought. Eventually, this developed into a new doctrine called AirLand Battle.[5] Today, we have a similar circumstance. Three recent conflicts—the Second Nagorno-Karabakh War (2020), the Israel–Hamas War (2021), and the ongoing Russia–Ukraine War (2022–)—provide us with the ability to learn from others' wars. There are those in our military dedicated to studying these wars in width, depth, and context, but more need to do so. We must use these conflicts to recognize the changes in warfare and reimagine how we fight.

"Are we ready for the next war?" This is the question our military and national security leaders must ask themselves every day. The US military is a full-spectrum force capable of meeting many challenges but, today, we find ourselves in a race with potential adversaries preparing to engage in large-scale combat operations. The measure of combat readiness is the degree to which a military unit or organization effectively conducts combat operations and achieves victory. Readiness is needed not only to win the next conflict but also to deter attacks. This ability is both intellectual and physical. The US spends more money than any other nation on defense. We have some of the most technologically advanced military systems, such as aircraft carriers, fifth-generation fighters, and the most powerful tanks. We assume our allies respect us and our enemies fear us—a notion we should reevaluate. All of this is an illusion if we cannot win wars. When was the last time the US won a war?

Over the past quarter-century, the US has exhibited a worrying lack of imagination and little prescience in deterring and winning wars. The attacks on September 11, 2001, against the World Trade Center and the Pentagon, and the subsequent retaliatory, long and indecisive wars in Iraq and Afghanistan, were significant failures. We must garner the courage to see these recent defeats for what they are: a failure of imagination. We need to do detailed and holistic after-action reviews to find out why we failed and learn from those lessons.

The US military is at an historic turning point. Technological convergence, in the synergy of micro-miniaturization, computing power, robotics, and sensors, is altering the methods of war. Of these, artificial intelligence (AI)[6] is speeding up this paradigm shift. Our AI today is still "narrow," and simple, but it enables a wide array of smart,

A Bradley Infantry Fighting Vehicle moves into position as helicopters fly overhead during a training exercise in Romania. US forces are deployed in Europe to join in the defense of the North Atlantic Treaty Organization. (US Army photo by Staff Sgt Malcolm Cohens-Ashley)

autonomous weapons that swim, drive, and fly throughout the battlespace. For many, the concept of ubiquitous sensors, robotic tanks, loitering munitions, unmanned combat aerial vehicles, long-range precision fires, and AI-enabled kill webs seem to have appeared overnight. This is not the case. These capabilities have evolved over the decades. Consider the Goliath (*Leichter Ladungsträger SdKfz. 302*), an electrically powered robotic weapon developed by the German Army during World War II; it was a small, tracked vehicle that contained 220 pounds of explosives and was nicknamed a "beetle tank" by Allied soldiers. Connected by wire to a joystick controller, one soldier could maneuver a Goliath toward a target and then command-detonate the mine with precise accuracy. During World War II, Germany manufactured 7,564 Goliaths and although it was not the first use of robots in war, it is an example of a robotic system deployed and used in combat.[7] Since then, significant progress has been made in the creation of robotic combat systems. Robotics and AI are clearly on the verge of revolutionizing military operations and ushering in a period of hybrid human–machine intelligent systems warfare.

The best way to prevent war is to be ready for it; if that fails, the goal is to win as quickly as possible. No one knows what will happen tomorrow, but not thinking about or planning for the future ensures you will be taken by surprise. Leaders who think and act in time are invaluable. Developing leaders with imagination and foresight is our most dire challenge. As vital as technology is to winning wars, human leadership remains paramount. Winning leadership in war requires great skills, imagination, and foresight. You engage your imagination when you read, think critically, ask pertinent questions, derive answers, and then test your conclusions. We develop foresight through reading, study, interaction, dialogue, wargaming, and red teaming. Develop your foresight today, not tomorrow.

The main aim of this book is to draw lessons and conclusions from the Second Nagorno-Karabakh War, the Israel–Hamas War, and the ongoing Russia–Ukraine War. These recent wars have been dominated by multidomain forces that used new means to influence combat operations. Faced with the need for change, the US Army has transformed its doctrine into a concept called Multidomain Operations (MDO).[8] This concept describes how the US Army, as part of the joint force (Army, Navy, Air Force, Marines, and Space Force) will counter and defeat near-peer adversaries capable of contesting the US in all domains (air, land, maritime, space, and cyberspace) in both competition and armed conflict. Eventually, this may evolve into a joint doctrine called Joint All Domain Operations for the entire US military.[9]

This is an unconventional book. Prepare for a wild ride. To impel your imagination and further your understanding of the changing methods of war, I introduce many of the changing methods of war with dramatic, **hypothetical** accounts. Imagine these as "thought experiments." To conventional thinkers, this may be jarring. A thought experiment is an imagined sequence of events that is used to illustrate or investigate the consequences of a given action or condition and attempts to solve a problem using the power of human imagination. I take inspiration for this approach from best-selling author and military futurist Peter Singer. In testimony before the US Congressional Armed Services Committee in February 2023, Singer emphasized the value of using hypothetical stories to inspire the imagination:

> A methodology for this cross of strategy and scenario is the deliberate blend of nonfiction with narrative communication techniques. Known as FICINT for "Fictional Intelligence" or "Useful Fiction," the goal is not to replace the traditional white paper, article, or memo, but to achieve a greater impact of research and analysis through sharing insights through the oldest communication technology of all: Story. The narrative is designed to allow a reader to visualize new trends, technologies, or threats, not just from altered perspectives, but in a format that the science of the brain shows is more likely to lead to both understanding and action. As such, the approach has been used by organizations that range from the US and NATO militaries to Fortune 500 companies.[10]

If *Next War* sparks your imagination, raises your awareness, and impels you to enter a dialogue with others about the changing methods of war, then it has accomplished its mission. A failure of imagination will get you killed. Embrace a bias for action. Ask difficult questions and then do not quit until you have answers. Use those answers to act in time. Who thinks, wins!

Acknowledgements

Every book is a journey, and it takes a team to win. Foremost, I want to thank my wife, Uncha. She makes everything possible and, without her love and support, I would never have been able to write this book.

I am honored and sincerely grateful to Senator Mike Rounds for taking the time to read *Next War* and write the foreword. His leadership in the United States Senate, and especially his work on the Armed Services Committee (SASC), is vital and an example of dedicated service to the nation. The SASC is empowered with legislative oversight of the nation's military, including the Department of Defense, military research and development, nuclear energy, benefits for members of the military, the Selective Service System, and other matters related to defense policy. As the Senator has said, "Our goal is to avoid war. Deterrence is the key." I could not agree more and his leadership in the SASC helps the US maintain that deterrence.

Next, I want to thank my friends who took the time to read, analyze, and improve this work. This list includes James Antal (Maj., US Marine Corps, Ret.), Mrs. Beth Antal, Daniel Adelstein (Lt Col., US Army, Ret.), Kevin Benson (Col., US Army, Ret.), Edward Braese (Sgt Maj., US Army, Ret.), Francis Fierko (Col., US Army, Ret.), Shawn Graves (Col., US Army, Ret.), Richard Jung (Col., US Army, Ret.), and Mrs. Carolyn Petracca. Your involvement was decisive; I thank you all.

I also want to express my appreciation to my publisher, Casemate Publishers, especially Ruth Sheppard, Isobel Fulton, Andy Wright, and Declan Ingram for their editing, support, and encouragement.

Finally, I wish to thank all the members of the US Armed Forces and allied officers who offered me encouragement and advice in finishing this effort. My purpose in life is to develop leaders and inspire service. You have helped me move closer to achieving that aim with your inspiration and commitment. To each one of you, I am sincerely grateful.

John Antal
May 17, 2023

This burning Russian tank was disabled by a Ukrainian small unmanned aerial system. Three recent conflicts demonstrate the future of military combat: the Second Nagorno-Karabakh War (2020); the Israel–Hamas War (2021); and the Russia–Ukraine War (2022–). The first was primarily won by robotic systems; the second by artificial intelligence; the third war is the largest conflict in Europe since 1945. (Ukrainian Army photo)

CHAPTER ONE

A Failure of Imagination

> It is much easier after the event to sort the relevant from the irrelevant signals. After the event, of course, a signal is always crystal clear; we can now see what disaster it was signaling since the disaster has occurred. But before the event, it is obscure and pregnant with conflicting meanings.[1]
> ROBERTA MORGAN WOHLSTETTER, ONE OF AMERICA'S MOST IMPORTANT HISTORIANS OF US MILITARY INTELLIGENCE

The only constant in leadership and war is change. Leaders need imagination to inspire foresight to visualize and prepare for the next fight. Developing your imagination concerning the changing methods of war is not an easy task. The great physicist Albert Einstein said, "Imagination is more important than knowledge. Knowledge is limited, whereas imagination embraces the entire world."[2] Einstein used thought experiments to explore and visualize complex subjects. Building on this concept, the following story is a thought experiment that has direct relevance to the changing methods of war today. It is a story of what could have happened in December 1941 to change the outcome of World War II. All it took to implement was thinking differently.

***

The only sound is the ticking of a clock. A calendar on the wall displays the date: November 25, 1941.

He knows that when experiencing a dilemma, it is helpful to change one variable, then reexamine the problem, but this idea is barbaric, terrible, inconceivable, brilliant, samurai, and noble. He recoils in horror at the suggestion, then embraces it with devotion. In an instant, he knows. Genda is right. Still …

Admiral Isoroku Yamamoto, Imperial Japan's best strategist and naval commander, shakes his head and sighs. He realizes a failure of imagination will cause the defeat of Imperial Japan. A failure of imagination will get his men killed and result in his personal disgrace. He is determined to make sure this does not happen.

His dilemma is agonizing. There is never enough striking power. His aim is to annihilate American naval power in the Pacific. Annihilation—not just damaging their fleet but destroying its capacity to wage naval warfare. Only complete destruction will do. To win a swift and decisive victory over the United States, the Japanese Imperial Fleet Strike Force must sink the American aircraft carriers, eliminate the capital ships berthed at Pearl Harbor, Hawaii, smash the dockyards and maintenance facilities, and burn the oil supplies.

The ships, especially the aircraft carriers and battleships of the US Navy, form the enemy's striking force in the Pacific. Sink these ships and it will take months or years for the Americans to respond. If the Japanese destroy the dockyards at Pearl Harbor, the Americans will not be able to repair damaged ships in the central Pacific. If Japanese carrier aircraft obliterate the oil and fuel depots at Pearl Harbor, it will deny the Americans the ability to fly and sail. Without fuel and repair shops, any surviving ships will be forced to leave Pearl Harbor. With Pearl Harbor neutralized, the Americans must sail to the dockyards and supply depots on America's west coast, and these surviving ships will run a gauntlet of attacks from Japanese submarines. Damaged ships, short on fuel and ammunition, will then be easy pickings for follow-on attacks.

But the current plan will not achieve this. The best the current plan will deliver is a stunning victory, but not the complete destruction of the American fleet. The numbers in the new report confirm this.

Anything less than the complete destruction of these three critical targets—ships, dockyards, and oil supplies—will deliver only a partial success. He must destroy all three target sets. With time, he knows the Americans will rearm, regroup, and attack with a force Japan will not be able to repulse. He feels confident his striking power is ample to wreak havoc for six months, but after that the overwhelming strength of an awakened, militarized, and industrialized America may grind the Japanese military into dust.

The weight of the decision he now faces is immense.

The Imperial Japanese Navy is the most powerful naval force in the Pacific Ocean, comprising 10 battleships, 10 aircraft carriers, 38 cruisers (heavy and light), 112 destroyers, 65 submarines, and many auxiliary warships of lesser size. A naval strike force of 51 ships, including six aircraft carriers—*Akagi*, *Kaga*, *Sōryū*, *Hiryū*, *Shōkaku*, and *Zuikaku*—will soon steam from Japan to attack the American fleet at Pearl Harbor. Still, Yamamoto knows success is a roll of the dice and he will only get one roll.

Yamamoto thinks about Genda's latest review of the plan to attack Pearl Harbor. Commander Minoru Genda is a brilliant naval officer, an operational genius and Yamamoto's trusted strategist. Based on guidance from Yamamoto, Genda planned the Pearl Harbor strike with meticulous genius, but is now having second thoughts. He has reevaluated the attack based on some disturbing information. His data proves the attack will not achieve Yamamoto's objective unless the Japanese take extreme measures. Genda is recommending dramatic changes to the plan, but the fleet will be underway in only two days. If Yamamoto accepts this new plan, there is little time to issue new orders and prepare.

With his usual thoroughness, Genda reported the highest dive-bombing hit rates in the past seven months of practice, by the Japanese Navy's best carrier pilots, is only 40 percent. The hit rate of the torpedo bombers is better, but only if the new Type 91 aerial torpedo modifications, which allow the torpedoes to operate in the shallow channel waters at Pearl Harbor, perform as expected.

The solution is now manifest. It is a matter of imagination.

Yamamoto sits in a spartan metal chair at a long table covered with maps in his ready room. He stares at the map of Pearl Harbor. Carefully drawn lines depict the approaches of his ships and the attack vectors his aircraft will fly to reach the American naval bastion. In his heart, Yamamoto realizes Genda is correct; they are all thinking in the "old way." They need to think differently—radically differently. Yamamoto desires a decisive solution to the American problem. Such a decision requires a unique and decisive approach.

But it is a terrible option. With only one chance, one roll of the dice, what should he do?

There is a knock on the door. "Enter," Yamamoto bellows.

It is Genda. The officer approaches the table and salutes.

"Are you convinced this is the only way?" Yamamoto asks, still studying the maps.

"Hai! Admiral, there is no other way. We will need three times the force, or at least six attacks, to accomplish our purpose in the old way."

"There is more than one path to get to the top of the mountain," Yamamoto replies.

"Hai!" Genda answers, recognizing Yamamoto quoting the philosopher and Samurai master, Miyamoto Musashi. "The only reason a warrior is alive is to fight, and the only reason a warrior fights is to win. Here, the path of life and death, victory and defeat, is clear."

"You realize what position this puts me in," Yamamoto answers as he looks up from the map and meets Genda's eyes with his.

Genda stands at attention, still holding his salute.

Yamamoto returns the salute, then points to the maps. "Stand at ease. We are less than two weeks away from launching the attack. How can we change our plan in such a short time?"

"It can be done," Genda replies. "It will take only six days to adjust the aircraft and we can do this while we are underway. With this new means, we will destroy the four American aircraft carriers, eight battleships, two heavy cruisers and the six light cruisers in the first wave. Conventional attacks will focus on attacking enemy airfields and destroying American planes on the ground. The second wave will target the dockyards and oil facilities. The third wave will involve conventional bombing and will hit any remaining targets."

"What about our losses? What do you predict?"

"We will lose 80 of our 353 aircraft through direct strikes," Genda replies. "Ten percent more if the enemy antiaircraft and their pursuit planes are alert … but I believe we will achieve surprise, so I estimate our losses at 107 aircraft."

"One hundred and seven," Yamamoto murmurs. "Is this the right way?"

"Yes, it is the only way to annihilate our enemy with one swift blow. It is a hard choice, I know, but these strikes will be like a Divine Wind that will blow the Americans from the Pacific."

Yamamoto pauses for a long moment. The distance from Japan to Pearl Harbor is about forty-one hundred miles. The distance from Pearl Harbor to the US naval ports in California is an additional 2,500 miles. Neutralizing Hawaii will force the US Navy to operate from ports in California. The Japanese homeland is 6,600 miles away. Such a victory will deliver a tremendous advantage to the Empire.

Yamamoto closes his eyes. The silence in the room is total, except for the ticking clock. Seconds pass in loud silence. The admiral says nothing. Then he opens his eyes and nods. "It is a terrible decision, but one we must make. War is not a game fought by rules. It is not a game at all. It is murder, and it is best done ruthlessly and completely. Our pilots must embrace this. I will not order them to do so against their will."

"You know our men. We will not lack for volunteers," Genda replies. "If you explain the reasons and ask them yourself, not a man will refuse. They are brimming with the spirit of Bushido. They know what is at stake and once we explain to them what must be done, they will know this is right."

"What is right?" Yamamoto retorts. "I fear you are correct. War does not determine who is right, it only determines who is left standing. In the end, Japan must be the one still standing."

"With your permission, admiral, I will give the orders," Genda replies. "Seventy of our planes in the first attack wave and ten in the second wave will form our Divine Wind. We will fit each aircraft with as much explosive as possible, so when they dive into American ships, dockyards and oil storage facilities, they will detonate in explosions that the Americans will hear in Washington, D.C."

Yamamoto nods in approval. "Put your new plan into motion. We will hit the Americans and destroy their power in the Pacific with one strike of the sword. We will use your 80 kamikaze aircraft to change the face of war."

"Hai!" Genda answers, snapping to attention.

Before dawn on the morning of December 7, 1941, the aircraft carriers in Yamamoto's strike group, led by Admiral Chūichi Nagumo, turn into the wind, and launch their aircraft. The first planes in the air are Zero fighters to protect the kamikazes in case the Americans are ready and have already launched fighters. Eighty kamikaze aircraft loaded with explosives and bombs follow these. As the first attack wave approaches Pearl Harbor, the codewords *Tora, Tora, Tora* echo over the Japanese radio net. These words signify the Japanese have achieved complete

Despite many warning signs, the US was not ready for the Japanese first strike in the Pacific that officially brought it into World War II. This is a photo of the battleship USS *Arizona* (BB39) burning after the Japanese attack on Pearl Harbor, December 7, 1941. Japanese dive bombers scored four hits and three near misses on and around *Arizona*. The supporting structure of the forward tripod mast has collapsed after the forward magazine exploded; 1,177 of the ship's officers and crewmen were killed. Today, the ship lies on the bottom of Pearl Harbor beneath the USS *Arizona* Memorial. (US Navy photo)

surprise. Brushing aside the few American planes in the air, the Zeroes clear the way for the kamikazes. In the next 30 minutes, in a horrible dance of death, 70 kamikazes dive on the American ships anchored at Pearl Harbor, with 67 hitting their mark. Sixty-seven ships blaze and explode as fires rip through their decks. Secondary explosions soon tear ships apart, and they sink to the bottom of the harbor. Japanese dive bombers and torpedo planes reinforce the kamikaze attacks by attacking the largest ships.

Dozens of Japanese aircraft also attack the American aircraft at Hickam and Wheeler airfields. The Japanese dive bombers smash the American planes parked wingtip to wingtip on the tarmacs. Other Japanese bombers devastate the American oil tanks, depots and dockyards. Tremendous explosions ignite thousands of gallons of fuel and oil. When the second wave of the Japanese attack force arrives, Pearl Harbor is so thick with black smoke that it darkens many of their targets.

Nevertheless, the second wave attacks and the kamikazes finish off any ships still above the water, while bombers hit the docks and oil fields for a second time.

The American Pacific fleet dies that day at Pearl Harbor. By what many see as a miracle, the US Navy's aircraft carriers are away from Hawaii on maneuvers and miss the Japanese attack.

The shock to the people back home in the USA is numbing. What kind of enemy would dive into our ships, intentionally committing suicide, to kill us? What kind of mad fanatics are we up against? How can we beat the Japanese without our Pacific Fleet?

***

Of course, this story never happened and is an exercise in counter-factual history. How the scenario would play out after such a fanatical assault is anyone's guess. Thankfully, the Imperial Japanese Navy did not use kamikazes in their attack on Pearl Harbor. Yamamoto never considered the need for such extreme measures and trusted in the traditional methods of warfare. The Japanese would only resort to kamikaze attacks in 1944, when their strategic military situation was dire, as they grasped for any means to strike back and delay the inevitable tide of defeat.

By 1944, American and Allied forces in the Pacific had defeated the Japanese in numerous battles at sea, in the air, and on blood-soaked islands scattered across the Pacific. The US Navy had four aircraft carriers at the time of the Pearl Harbor attack. By 1945, it alone had 44 to support the invasion of Okinawa, while the Japanese Navy had no operational aircraft carriers. Unable to match the Americans in ships, aircraft, land power and every other means of modern warfare, the Japanese ordered their pilots to commit suicide and crash their planes into American ships. In October 1944, at the Battle of Leyte Gulf, 55 Japanese kamikazes sank one US escort carrier and five other ships, and damaged six aircraft carriers and 40 other ships. In desperation, many Japanese pilots became deadly, human-guided precision weapons. Later, in the Battle of Okinawa, which lasted from April 1 to June 22, 1945, kamikazes hit 130 US and Allied ships, sinking two escort carriers and three destroyers. These furious, sacrificial efforts, however, were not enough to stem the tide of American victory, but they ignited alarm in the Allied force. By the end of World War II on August 12, 1945, the US Navy comprised an overwhelming force of 6,768 ships, including dozens of aircraft carriers and battleships.[3]

But what if Yamamoto's forces had conducted the kamikaze strike strategy at Pearl Harbor in 1941, when the US Navy was much smaller and unprepared for such a ferocious assault? What if the Japanese had realized they had to play their one roll of the dice differently? How would history have turned out if Yamamoto had asked his pilots to be kamikazes?

Warfare is very different today. In today's multidomain battlespace, commanders do not have to order human pilots to fly kamikaze missions. Robotic systems can accomplish this role with precision and with no loss of friendly lives. The kamikaze has transformed into a weapon that is unmanned, precise, stealthy, and difficult to defeat. Considering the changing methods of warfare, these robotic systems are relatively new weapons. The next Pearl Harbor attack will most likely involve long-range precision fires: missiles, unmanned combat aerial vehicles, and loitering munitions. These new technologies are making a tremendous impact in the modern battlespace. It is vital for us to imagine how to use these weapons to win the next war.

In modern warfare, a failure of imagination will get you killed. Reinforce this failure with hubris, an attitude that you do not need to adapt since no enemy can match you, and the nation's survival is at risk. War is something that happens "over there," to other people. The US is a superpower. It has never known defeat. As a people, Americans cannot imagine losing. Yes, there were those long wars in Vietnam, Iraq, and Afghanistan, but do those count? Who could defeat the US? After all, it has the best military in the world, right?

The purpose of America's military power is to deter conflict and win the nation's wars. Its opponents have studied its methods and failures. The US cannot afford to be unprepared in a fight against a peer enemy. If it does not adapt its thinking and act in time, even wars against lesser powers could be disastrous. There is an urgent need to think differently.

Recent wars offer numerous lessons concerning thinking and imagination. In September 2020, the Azerbaijanis unleashed a war against the Armenians in Nagorno-Karabakh. This war must be studied in every US military school. Prior to 2020, few thought about Azerbaijan as a military power, yet they conducted a skilled, joint, multidomain campaign against their opponents. They used new technologies not common in the US military. Fighting against an enemy that held prepared mountain defenses, the Azerbaijanis executed a 44-day military campaign that ended in a decisive victory. The campaign was not flawless, but the Azerbaijanis "got more right" than their opponents, and that is all that counts in war. Azerbaijan leveraged drones, electronic warfare, and cyberwar to overturn the defender's traditional advantages. The combination of "shock drones"[4] and a sophisticated sensor and battlespace coordination effort by the Azerbaijanis and their Turkish allies destroyed the Armenian air-defense network and provided Azerbaijan with air dominance for the rest of the war.

Instead of studying this important conflict, the lessons of the Second Nagorno-Karabakh War are largely unknown in the US military. This is unfortunate as this war was the first in history to be won by leveraging robotic systems and it pre-shadowed the course of combat in the Russia–Ukraine War.[5] As the Secretary of the US Army, Christine Wormuth, stated at the Association of the US Army conference on October 11, 2021: "I'm not convinced that we have fully thought our way through all of the challenges we may face on the future high-end battlefield if deterrence fails. We need to look harder at key cases, such as the Nagorno-Karabakh War between Armenia and Azerbaijan ... Perhaps hardest of all, these changes must be made swiftly."[6]

The second war to examine occurred in the spring of 2021. Operation *Guardian of the Walls*, the name used by the Israel Defense Forces (IDF) for the 2021 Israel–Hamas War, lasted only 11 days, but this conflict showed the possibilities of wars to come. The IDF called this the "first artificial intelligence war,"[7] as AI was the key element in their success. Hamas launched 4,360 rockets at Israeli cities and towns, many of which were knocked out of the sky by Israeli Iron Dome anti-missile rockets. The Israeli anti-missile defense was impressive but, to end the war, the IDF had to take the fight to the enemy. Hamas fighters were hiding among the people and Israel's dilemma was to separate combatants from non-combatants in a dense urban battlespace. Israeli sensors collected years of data on their enemies from all sources, centralized this information into a multidomain sensor database, and accessed it in real-time to generate multidomain targeting information. Sensors input data continuously and in real-time to update a common operational picture that provided the IDF with a transparent view of their opponents. The IDF also used AI-enabled drone swarms for sensing and striking. The AI generated a super-fast kill chain that enabled the IDF to eliminate enemy fighters and destroy Hamas rocket launchers while minimizing civilian casualties within the city of Gaza.

The third war that must be studied is the ongoing Russia–Ukraine War. In late 2021, the idea that Russia would attack Ukraine and begin the deadliest war in Europe since 1945 seemed possible but unlikely.[8] By mid-January 2022, the intelligence view changed and the US seemed certain the Russians would attack Ukraine, but American diplomacy did not deter the Russians. Just before the invasion, the US Chairman of the Joint Chiefs of Staff, Gen. Mark Milley, told lawmakers that "Kyiv could fall within 72 hours if a full-scale Russian invasion of Ukraine takes place."[9] When Russia invaded on February 24, 2022, US intelligence agencies, calculating the size and capabilities of the opposing forces, predicted the Russians would take Ukraine in two or three days. Apparently, the Russians believed the same logic and thought the Ukrainians would greet them with bread, salt, and flowers, traditional Russian gifts for greeting important guests. Instead, the Ukrainians greeted the invaders with Molotov cocktails, Javelin anti-tank guided missiles (ATGMs), and Kalashnikov assault rifles. Ukraine's will to resist surprised the world. It is difficult to measure an army's willingness to fight; the courage and determination to adapt, improvise, and overcome is something we can only guess at until proven in battle. Second only to the Ukrainians' will to fight is their ability to think and create solutions as they confront the ever-changing methods of warfare. The Russians smashed into them like a tidal wave on February 24, 2022, and the Ukrainians learned to swim with the tide. Fighting back fiercely against the invasion and adapting rapidly in the chaos of the first weeks of battle, the Ukrainians beat the odds. This ongoing conflict is also turning out to be the most technological war in history.

These wars show that a one-punch boxer will lose to a skilled opponent who can use both arms. The contest is even more one-sided if the opponent is a skilled mixed martial artist who uses their torso, head, legs, and arms to win. The modern multidomain battlespace is like this mixed martial artist. Armed forces that can work together, coordinate, and synchronize operations on land, at sea, in the air, in space, and in cyberspace, as well as influence the electromagnetic, informational, and human dimensions, will defeat adversaries who only operate in one or two domains. Multidomain capabilities, that are now possessed by even third-tier military powers, have made the battlefield—the modern term is "battlespace" to depict the multidomain nature of the modern battle areas—transparent. Multidomain sensor networks use sensors from the muddy ground to outer space to reveal targets in the battlespace. Creating an unblinking eye that identifies, locates, and tracks targets in a congested battlespace is not a simple task. It takes sophisticated systems, purpose, and planning to reveal the enemy in this new battlespace. Couple ubiquitous sensors with long-range precision fires, that can hit and destroy targets at extreme ranges, and you grasp the crux of the issue. Indeed, in Nagorno-Karabakh in 2020, Gaza in 2021, and Ukraine from 2022, the combined effort of sensors, drones, and long-range precision fires shaped the conduct of these wars.

This does not mean the traditional means of war are not vital, but it does mean we must reimagine how to fight and develop new combinations of combat power. Since World War II, the US military has enjoyed air dominance due to its superior air forces, but this is no longer assured. "The threat has changed. Our adversaries, large and small, now integrate Intelligence, Surveillance and Reconnaissance (ISR) sensors, especially UAS [unmanned aerial systems], with long-range precision fires. For US forces, this is the end of guaranteed air superiority."[10] War is now a matter of "finders," versus "hiders," and "strikers" versus "shielders." Imagine a peer fight against China where US forces rush to defend Taiwan. As with the Japanese in World War II, American and Allied forces are "in the way" of Chinese ambitions. How will our forces survive and win when sensors see everything and long-range precision fires target anything seen? Do our forces have the skill, training, and equipment to survive an enemy first strike? Will our forces adapt to the ever-accelerating tempo of war? Can we execute Mission Command[11] in a degraded communications environment? How will our command posts survive? Will commanders see and understand what is happening in the battlespace to plan, decide, and act in time?

An example of a cutting-edge unmanned combat aerial vehicle is the Gray Eagle, a medium-altitude, long-endurance unmanned aircraft system. The MQ-1C Gray Eagle is manufactured for the US Army by General Atomics Aeronautical Systems. It was introduced in 2009 and is capable of transporting either 8 x AIM-92 Stinger air-to-air missiles or 4 x AGM-114 Hellfire anti-tank guided missiles. (Photo by US Army)

How do we answer these questions? First, we must visualize the problem. You can do this by creating your own thought experiments and war gaming new solutions to pressing problems. To think differently, visualize how yesterday's contests relate to today's challenges.

> The first and most important lesson is to understand and internalize the idea that we stand at a precipice of change, where our time-honored success and the ideas, concepts, doctrine, equipment, training, and personnel that achieved them probably are insufficient to achieve successes in the near term, and certainly are, if not revised or re-assessed, insufficient in the mid- to long-terms.[12]

As part of this first step, it is important to have clear definitions, as all understanding starts with definition. Here are three: The first is "Battleshock." Battleshock is the operational, informational, and organizational paralysis induced by the rapid convergence of key disrupters in the battlespace. Battleshock occurs when the tempo of operations is so fast, and the multidomain means so overwhelming, that the enemy cannot think, decide, and act in time. The second is "masking." Masking is the full spectrum, multidomain effort to deceive enemy sensors and disrupt enemy targeting. It requires commanders take action to deceive and disrupt in everything their units do. Masking is essential to survive and win in the modern battlespace and should be a principle of war in the 21st century. The third is "Mobile Striking Power," the offensive capacity of any military system, unit, or force to generate offensive action to move across the battlespace and disable or destroy the enemy. Mobile striking power is essential to offensive action. Although defense may be the strongest form of war, nations do not win wars by defending. Only offensive action brings victory.

Creating battleshock accelerates decisive military operations. Except for "Weapons of Mass Destruction," war is not a matter of annihilation, but of making the enemy run and then finishing them. Attacking the enemy's ability to command and control disrupts his striking power. Modern warfare is, therefore, a shot to the head, not a bullet in the arm. Striking power is the gun that takes the shot. Masking protects you from the enemy's striking power. Striking power combined with masking can generate the battleshock to overwhelm your opponent. Let us now investigate ways to accomplish this.

The Top Nine Disrupters of Modern Warfare

Disruptions are frightening. We live in an era of exponential change and constant disruption. Political, social, technological, and biological disruptions have marked the first two decades of the 21st century, and the years ahead promise more of the same, but with an increasing tempo. There is a tendency to perceive new military technologies as "game changers" in warfare. Although new weapons bring new possibilities, their impact is often over-hyped. The term "game changer" should be

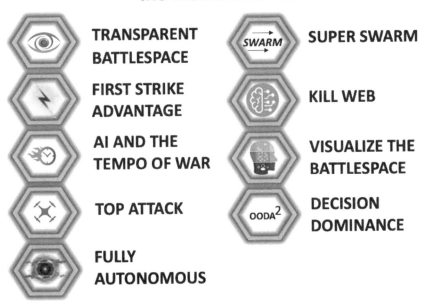

retired—immediately—as it is fuzzy, not focused, thinking. Drones, for instance, have not done away with the vital need for infantry, tanks, and artillery. Infantry is not obsolete because of the machine gun, nor are tanks a thing of the past because of ATGMs. A careful review of history shows new technologies by themselves do not change warfare, but novel applications, concepts, and combinations of those technologies do.

These top disrupters, derived from a deep study of recent wars, specifically the Second Nagorno-Karabakh War, the Israel–Hamas War, and the ongoing Russia–Ukraine War are: the transparent battlespace; the first strike advantage; artificial intelligence and the accelerating tempo of war; top attack; the shift from semi-autonomous to fully autonomous weapons; super swarms; the transition from a kill chain to a kill web; the ability of commanders to visualize the battlespace; and decision dominance. These are the essential elements of the next war.

CHAPTER TWO

The Transparent Battlespace

> The next war is highly lethal ... with sensors everywhere, the probability of being seen is very high. And as always, if you can be seen, you will be hit. And you'll be hit fast, with precision or dumb munitions, but either way you'll be dead.[1]
>
> GENERAL MARK A. MILLEY

A Russian electronic-warfare (EW) soldier sits in the operations station of a LEER-3 (РБ-341В Леер-3) EW system inside a KamAZ-5330 truck. The operator looks at his computer screen as an icon changes color, alerting him to important new information.

"We have them," he announces to the officer sitting in a chair next to him.

"The drone has identified their coordinates?" The officer, a major in the Russian EW forces, asks.

"*Da*," the Russian sergeant answers. "Confirmed."

"Now is it time to see where their commander is," the officer replies.

A report from Russian intelligence is in front of him. It provides him the telephone number he needs. He then reaches for a cell phone and dials the number.

On the other end of the call, the phone rings. A woman's voice answers.

"Hello, this is Mrs. Dacko."

"Mrs. Dacko, I am Captain Chornavil from the Ukrainian Army," the Russian major says in perfect Ukrainian. "I am extremely sad to inform you that your son, Major Dacko, is gravely wounded."

"No! My God, my God," the woman replies. "How? Where?"

"We don't know all the details right now, as I am at the headquarters, not the hospital. Maybe you can call him? He may still have his cell phone with him. Again, I am sorry for this, Mrs. Dacko. I must go now. I am sorry to tell you this bad news. God bless you."

The Russian major ends the call.

"Do you think it will work?" The Russian EW operator asks.

"We will know in a few minutes," the major replies.

Major Sergey Ivanovitch Petrov had trained for this mission for weeks. He trusts his equipment. The LEER-3 system, sharing data with several Orlan drones, can discover a unit's location based on its electronic signatures, such as global-positioning satellite navigation and cell phone communications.

The sergeant's computer screen flashes another indicator.

"It worked," the sergeant announces. "I located the ping of the cellphone. She is talking to her son."

"You can always count on a mother's love." Petrov says with a smile. "Send the coordinates to the artillery group."

The Russian sergeant quickly complies and sends an encrypted digital message.

Many miles away, in an open field, an unsuspecting Ukrainian task force is arrayed in an assembly area near the Ukrainian town of Zelenopillia. The ad hoc unit is composed of one mechanized infantry company in eight-wheeled BTR armored personnel carriers, a few BMP infantry fighting vehicles, a battery of self-propelled guns, three T-64 tanks and a dozen trucks and support vehicles. Located 16 kilometers (10 miles) behind the front line and beyond the range of the Russian separatists' short-range artillery, the Ukrainian commander feels safe.

Camped near the main road, the Ukrainian soldiers plan, perform maintenance on vehicles, resupply, and rest. When the sun rises, they will assemble, move to the front, and engage the separatists. Few of the soldiers on guard that night give much attention to the buzzing sound overhead. Russian and Ukrainian drones have flown over them in the past two days, without incident. Yesterday, just before nightfall, the Ukrainians downed a Russian Orlan-10 drone. Since they were out of range of the separatist guns, the Ukrainians disregarded the drones.

Last week, the Ukrainian Army launched a major operation to restore control of the Ukrainian border. This offensive disorganized the separatists, dislocated their defenses, and forced the enemy to retreat. Disheartened, and faced with continuous Ukrainian attacks, the separatists seem to have lost their fighting spirit. The Ukrainian commander did not expect much of a fight tomorrow. He anticipated it would be a simple drive to the border and then the occupation of a new defensive line.

Near the town of Zelenopillia, Ukrainian forces and equipment are assembling. At dusk, more armored vehicles and trucks show up. Instead of positioning these new troops and vehicles in the nearby woods, and have them move around in the dark, the Ukrainian leader chooses to set them in an open field for the night. Although the danger appears small, as he is sure he is out of range of the enemy's artillery, he does not want any of his men to be struck by a vehicle at night. Sergeants are told to line up their machines like a garrison motor pool. Supply trucks are unloaded, and boxes of ammunition are stacked in the open field.

Ukrainian soldiers sleep wherever they can. Some set up small tents or just plop down with blankets and sleeping bags on top of their armored vehicles. No foxholes or trenches are dug. Six officers huddle in a tent that serves as the battalion command post. A small stove inside the tent offers warmth and men brew tea as the officers enjoy the heat of the stove.

Russian TOS-1A multiple rocket launchers practice firing at a training range in Russia. TOS stands for the Russian equivalent of "heavy flame thrower." The TOS-1A launches thermobaric rockets with fuel–air explosive warheads similar to napalm. A rocket barrage by one TOS-1 system will destroy everything within the 200 × 300-meter blast zone. Systems like this were used by Russian forces to strike Ukrainians at Zelenopillia and are in action today in the Russia–Ukraine War. (Wikimedia Commons)

"Tomorrow we will attack the separatists as they are eating their breakfast," the commander announces to the men standing around the stove. "We'll use our tanks and BMPs to cut off the rebels' resupply route. Then we'll secure the Ukrainian border. Our border."

His men respond in approval. "We will teach them how Ukrainians fight," a captain adds.

"Good," the commander replies. "We are ready for tomorrow's battle."

He is dead wrong.

At 4:30 am, just before dawn, battle comes to the Ukrainians. Russian jammers disrupt Ukrainian electronic communications, rendering their radios useless. Salvos of thermobaric rockets explode in the assembly area. As these bombs detonate, the open field near Zelenopillia becomes hell on earth. In all the noise and confusion, it is impossible to count the explosions. There is nowhere to hide, nowhere to run, and no way to fight back.

***

The dramatization above shows how "pinpoint propaganda"[2] and ISR (intelligence, surveillance, and reconnaissance) drones can locate targets for long-range fires.

The actual attack in this story occurred on July 11, 2014, near Zelenopillia, in the Donbas, Ukraine.[3] According to a 2015 study by Dr. Phillip A. Karber, at Zelenopillia, "Two Ukrainian mechanized battalions were virtually wiped out with the combined effects of top-attack munitions and thermobaric warheads."[4] The Russian fire strike lasted only three minutes. The 1st Battalion of the 79th Mykolaiv Airmobile Brigade took heavy casualties during the rocket onslaught. When the fire strike finished, the Ukrainians lost 37 killed and more than a hundred wounded. A colonel from the Ukrainian border guards was among the dead. Many of the wounded in very grave condition, burned by the thermobaric explosions, would not survive. It was a disaster.[5]

The strike at Zelenopillia in 2014 was the first active entry into the Donbas fighting by the Russian Army. Before this attack, the Russians had supported the separatists but had not entered the fighting in a major way. The Ukrainians were not ready for Russian intervention. Now the gloves were off. In the next six weeks, the Russians conducted 53 fire strikes against Ukrainian units, with horrific losses for the Ukrainians, and consequently changed the course of the war. Zelenopillia has three different accounts—one provided by US analysts, one by the Ukrainians, and the third by the Russians. But the central idea of each tale stays the same: drones were used by the Russians to find the Ukrainians, who were then carefully targeted and destroyed in a coordinated fire strike. In today's battlespace, networked surveillance systems can locate targets across the depth of the battlespace in the optical, thermal, electronic, and acoustic realms. The critical lesson from Zelenopillia, and the ongoing Russia–Ukraine War, is that there are no sanctuaries.

After an investigation of the Russian fire strike at Zelenopillia by the Ukrainian general staff, the task-force commander was relieved of command for incompetence and for not ordering his soldiers to dig in, disperse, and occupy the forests instead of making camp in an open field. They learned the vulnerabilities of soldiers using cell phones. In this modern interconnected world, people are so addicted to their information devices that being without a smartphone has its own term—"nomophobia"[6]—defined as the fear of being without a mobile phone. Since smartphones are easily tracked, a major challenge to military leaders today is to ensure none go to war, something that few, if any, military units in the world have been able to enforce. The Ukrainians learned vital lessons from the Zelenopillia disaster that would change how they fight in the years ahead. They learned the power of the Russian reconnaissance–strike complex and swore the soldiers who perished at Zelenopillia did not die in vain but would become poignant lessons for future Ukrainian fighters, and examples of the transparent battlespace of modern warfare.

Mobile phones easily reveal friendly locations. A 2018 article by Colonel Liam Collins in *Army Magazine* acknowledged this vulnerability: "It's a nightmare scenario. An enemy unmanned aerial vehicle monitors the cellphone signals of troops below, identifies their location and sends the coordinates to a headquarters, which launches an artillery strike against the unsuspecting troops."[7] Locating cellphone signals was how the Russian command explained the slaughter of anywhere from 90 to 200 of their own soldiers at Makiivka on January 1, 2023. Reports on the exact number of casualties vary, but the losses in this single attack on a Russian Army barracks, within range of Ukrainian long-range precision fires, were dramatic. Attempting to defer blame for this catastrophe, Russian Lt. Gen. Sergei Sevryukov reported: "But is it already clear that the main reason for what happened was the switching on and mass use by personnel—contrary to the ban in place—of cell phones in a strike zone accessible to enemy weapons. This factor allowed the enemy to track and determine the coordinates."[8] The lesson is clear: Remember Makiivka. Cell phones in the battlespace reveal your position and will get you killed.

Seeing enemy forces in the battlespace is the most revolutionary disrupter to traditional methods of warfare. In the past, the enemy had to be seen with the

The battlespace is now transparent. Modern sensors can see targets in the optical, thermal, electronic, and acoustic realms. A few exquisite sensors can even identify targets using quantum sensing. This photo is an image from an airborne sensor that identified the electronic signatures of a US Army brigade at the National Training Center, Fort Irwin, California, in 2020. Sensors that provide similar images will be used by America's peer enemies to see and target US forces. (US Army photo by Scott Woodward)

human eye, heard with human ears, and visualized with the use of analog maps and terrain models. Seeing at night and in extreme weather was difficult. Effective armies took extraordinary means to mask their locations, such as the use of smoke, nighttime, storms, and the natural obscuration of forest, mountains, and urban terrain. In the past, knowing what was over the next hill, or around the corner in a city fight, required human reconnaissance, but today sensors are replacing human-in-the-loop cognitive processing.

These sensors are often located above you, looking down with an unblinking eye. When formed into a sensor network, they reveal nearly everything and make it very difficult to hide in the modern battlespace. Inexpensive and effective sensor networks, comprising ground sensors, drones, radar, aircraft, and multispectral imaging satellites, form the transparent battlespace. A standard visual sensor collects red, green, and blue wavelengths of light. Multispectral sensors collect these visible wavelengths, as well as wavelengths that fall outside the visible spectrum, including near-infrared radiation, short-wave infrared radiation, and others. If the enemy cares enough to position the right sensors, and our peer adversaries have such sensors, then you will be observed. Unless you take measures to deceive these sensors, you will be targeted.

The battlespace is now "naked," day and night, and everything in it can be seen by sensors and targeted. A layer of sensors can stream real-time to reveal the battlespace, confirm battle damage, and, in consonance with information from other sources, render the battlefield transparent. An array of standard and multispectral sensors can see the battlespace in five main areas of observation, or signatures: optical, thermal, electronic, acoustic, and quantum. Let us review each one of these and discuss ways to mask against them.

Optical Signature

We are familiar with the optical spectrum as it is what we can see with our own eyes. Aiding the human eye with optical devices, such as the telescope, became common practice after Galileo perfected the telescope in AD 1609—an example of the military use of commercial technology. Binoculars were first used by military commanders around 1854 and became common during World War I. Much has changed since then. Today, unmanned combat aerial vehicle (UCAV) systems, such as the Bayraktar TB2, view the battlespace with excellent high-definition optical sensors. Camouflage, the art of blending in with your surroundings, is the solution to deceive these sensors, but few armies today are adept at it. Hiding in plain sight provides protection and offers the ability to surprise the enemy. Nature embraces camouflage and a predominant rule of the animal kingdom is not to be seen unless you want to be. The tiger, for example, has patterned fur that blends with the color of its habitat. The chameleon changes colors as it senses changes in its environment.

Humans have embraced camouflage since the beginning. Armies only wore colorful uniforms when it became necessary for leaders to better command their actions and for forces to identify friend from foe. World War I made modern camouflage necessary for survival. Colorful uniforms faded from use in the trenches because of the killing power of machine guns, rapid-firing artillery, and aircraft. After taking terrible losses to German firepower, and starting the war in brightly colored uniforms, the French used their imagination to hide in plain sight and were the first combatants in the Great War to form special camouflage units. They enlisted zoologists, artists, and theater set designers to develop techniques to obscure critical positions, equipment, vehicles, and command posts. The people who did this work were called "camoufleurs," and their name became the inspiration for the term "camouflage."[9]

Disruptive camouflage—camouflage that works by breaking up the outlines of a soldier or military vehicle with countershading or a contrasting pattern—soon became common, especially in the Royal Navy, which painted its ships in a disruptive pattern called "Dazzle" in an attempt to reduce the number of British merchant ships lost to German submarines.[10] They painted a dazzle design on over 2,300 British warships and merchant vessels during World War I. Hugh Cott, a British zoologist who worked as a camouflage expert for the British Army in World War II, explained the purpose of disruptive camouflage: "The function of a disruptive pattern is to prevent, or to delay as long as possible, the first recognition of an object by sight … irregular patches of contrasted colours and tones … tend to catch the eye of the observer and to draw his attention away from the shape which bears them."[11] Cott emphasized the essential factor for effective camouflage was to disguise the surface continuity and contour of the thing you wished to disguise and must conceal at ranges at which the object was most likely to be detected. Other navies, such as the US, where it was called "razzle dazzle," used the design. After World War II, dazzle camouflage was abandoned as the capabilities of ship rangefinders, optical sights, and aircraft reduced its effectiveness.

The purpose of camouflage is concealment and deception. Armies that fought with limited air power, such as the German Army in the later years of World War II (1943–45), and the communist Vietnamese forces in the First Indochina War (1946–54), and the Vietnam War (1955–75), became experts at camouflage out of sheer necessity. In both cases, the Germans and the Vietnamese adopted strict camouflage discipline and techniques to conceal their forces from observation by enemy aircraft to avoid being strafed and bombed. For the Vietnamese, who were outmatched by the Americans in technology, camouflage was life. They knew they must always camouflage their forces to survive, communicate, move, and attack.

Today, the US Army teaches camouflage, cover and concealment techniques that involve hiding, blending, disguising, disrupting, and decoying. Hiding puts a barrier, such as a camouflage net, between the observer and the target. Blending alters the target's appearance to become part of the background. Disguising applies materials

A camouflaged French military vehicle enters the Joint Multinational Training Center, at Hohenfehls, Germany, on January 24, 2020. The type of camouflage seen in this photo is manufactured and designed to break up the physical appearance of the vehicle and reduce the vehicle's thermal and infrared signature. (US Army National Guard photo by Sgt Fiona Berndt)

to mislead the observer into thinking the target is something else. Disrupting alters the target's regular pattern and characteristics. Camouflage discipline is weak in the US Army and only slightly better in the US Marine Corps. The US Air Force (except for stealth aircraft) and the US Navy (except for submarines) have enormous challenges. Camouflaging an aircraft carrier is impossible; hiding an airfield is much the same.

To be excellent at the art of camouflage, hiding, blending, disguising, disrupting, and decoying must become second nature to every soldier and marine. If advanced camouflage systems are not available, local materials and natural camouflage should be used. In past wars, the combatants produced most camouflage materials ad hoc, using whatever was available. In industrial-age warfare, more sophisticated camouflage uniforms, painted equipment, and netting were developed and deployed. Today's camouflage, however, must hide combatants from sophisticated optical sensors and high-definition cameras. Indeed, the old means are still vital, but we must field advanced multispectral camouflage systems to mask in the optical spectrum. "We have got to get our hands around fighting under continuous observation," Gen. James Rainey, the commanding general of United States Army Futures Command, said on October 12, 2022, at the Association of the US Army's annual convention in Washington, D.C. "You're going to have to figure out how to fight when the enemy can see you. You're not going to be able to pile up things; you're not going to be able to build TOCs (Tactical Operation Centers)."[12]

Camouflage requires equipment, tactics, techniques, and procedures (TTP), and leaders to enforce camouflage discipline. Inadequate equipment, such as antiquated camouflage netting, does not hide you from the unblinking eyes of the latest drones as their sophisticated electro-optical cameras are too discerning. New multispectral camouflage netting and systems are required. Poor TTP, such as not camouflaging generators and support equipment, or not camouflaging at all, gives away friendly positions. Ineffective camouflage that cannot hide personnel and equipment in a transparent battlespace is a recipe for casualties. Leaders who do not understand the need to camouflage all the time, and take no action to implement camouflage discipline, fail the warfighters they lead.

In early December 2022, the Ukrainians provided an example of the transparent battlespace. Satellite imagery of several Russian air bases inside Russia, indicated the Russian Air Force—the VVS (Военно-воздушные силы России)—was preparing for massive missile strikes against Ukraine. Commercial satellite imagery, available on military-oriented *YouTube* channels, depicted the VVS aircraft and missile-storage areas in high-definition photographs and real-time video feeds. If these YouTubers accessed this open-source information, then imagine what was available to the Ukrainians who had NATO and US intelligence satellites and airborne-warning-and-control system aircraft support. The Russians could not hide their Tu95 and Tu160 bombers on the airstrips of Engels-2 and Dyagilevo air bases. In the early hours of December 5, 2022, the Ukrainians used reconnaissance data to launch a preemptive precision strike against aircraft parked on the undefended runways.

These strikes, most likely made by repurposed Ukrainian Soviet-era Tupolev Tu-141 Strizh reconnaissance drones, damaged a few of the VVS bombers and killed and injured personnel on the ground. The Russian Ministry of Defense reported Ukrainian drones penetrated Russian airspace to strike two air bases in south-central Russia, at Ryazan and Saratov. These air bases were deep inside Russia, with Engels, near Saratov, more than 370 miles from the Ukrainian border. Dyagilevo, near the Russian city of Ryazan, is only 122 miles southeast of Moscow.

The Ukrainians did not immediately take credit for the attack but inferred they were responsible. "If something is launched into other countries' airspace," said Mykhailo Podolyak, an advisor to Ukrainian President Zelensky, "sooner or later unknown flying objects will return to the departure point."[13] Seeing deep behind enemy lines, the Ukrainians were able to use this drone raid to create operational advantages, similar in effect as the American Doolittle raid had against Japan in 1942. Both operations produced little damage, but they shifted the reaction of the nations targeted. With the Ukrainian attack on Russia bomber bases, the Russians shifted air defenses, that were otherwise needed closer to Ukraine, to defend airfields they previously considered "in sanctuary." As seen by the Ukrainian strikes on Ryazan and Saratov, modern, multidomain sensors make it extremely difficult to mask effectively to avoid being targeted by long-range precision fires and unmanned aerial systems.

Thermal Signature

As in the optical spectrum, masking thermal signatures demands new equipment and new thermal camouflage. We do not design armored vehicles and trucks with masking in mind to reduce the system's thermal signature. The M1A2 SEPv3 (System-Enhanced Package Version Three) Abrams, the most modern tank in the US Army's inventory, has a powerful gas-turbine engine that, when operating, is easy to identify with thermal sensors. Thermal camouflage is available that can reduce thermal signatures, such as the Saab CoolCam Mobile Camouflage System, but it is expensive and only partially effective. UCAVs, with state-of-the-art optical and thermal sensors, will spot and target unmasked vehicles. Thermal camouflage is pricey, and may only deceive some of the enemy's sensors, but losing soldiers and equipment is devastating. If given a choice between expensive and devastating, which would you pick?

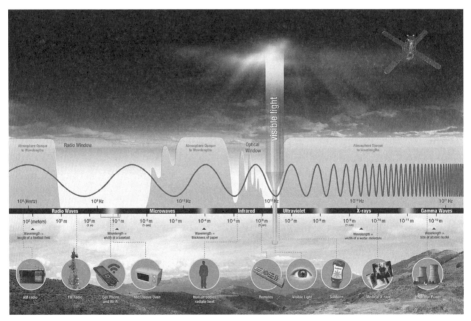

Electromagnetic energy travels in waves and spans a broad spectrum from very long radio waves to very short gamma rays. A radio detects a different portion of the spectrum, and an x-ray machine uses yet another portion. The human eye can only detect a small portion of the electromagnetic spectrum called visible light. (NASA photo)

Electro-optical and thermal-targeting systems have advanced significantly over the past 20 years. Satellites, unmanned systems, aircraft, armored vehicles, and individual soldiers now have access to an array of thermal-imaging devices. Thermal cameras on these systems capture infrared light, which is invisible to the human eye, and detect the differences in the infrared emissions of objects and their background. Anything that is hotter or colder than its surroundings emits a thermal signature. Thermal imagery, which is emissive rather than reflective, is very difficult to mask. Traditional camouflage is not enough.

> In 1958, US Army scientist John Johnson developed a criterion for measuring visual thresholds, which includes thermal vision, by assuming range performance is proportional to image quality. Labeled as "Johnson's Criteria," it listed the minimum resolution to detect, orient, recognize, and identify targets in terms of the size of the pixels produced by the imaging device. These measurements give a 50-percent probability of an observer discriminating an object to the specified level with precise specifications for detection, orientation, recognition, and identification.[14] A newer criterion for measurement, the Targeting Task Performance metric, developed in 2004, is used today to determine the best range of electro-optical sensors and thermal-sensor capabilities. In a 1990 study, the US Army defined target recognition as choosing an object from a known confusion set by differentiating it from the possible alternatives.

Thermal military sights were common on US systems in the 1980s and provided a stunning advantage on the battlefield. Until recently, many competing systems did not have thermal-sight systems. Today, the Russian T90M tank (Прорыв-3, Breakthrough-3), and the Chinese Type 99, have third-generation thermal-imaging sights. Even the Iranian Karrar (Striker) main battle tank has modern aiming components and fire-control systems with thermal sights. We must assume adversaries will see us as well as we can see them. In short, the US no longer has a singular advantage in thermal imaging and must act to reduce the thermal signature of its equipment and personnel.

Masking military equipment to deceive enemy sensors in the thermal spectrum is a tremendous challenge. Gaining thermal "invisibility" would be an amazing tactical advantage, even if the system is off and at rest, but this goal has proven elusive. Russian equipment designers bragged that their new systems reduced thermal signatures, making their tanks and armored vehicles difficult to see and target. Vyacheslav Khalitov, the deputy chief executive officer of the *Uralvagonzavod* scientific-industrial complex, reported that the Russian T14 Armata defeats some of the West's advantages in thermal sights. "New Russian tanks would include stealth technology," Khalitov said in a 2017 interview in *Russia Beyond*. "Radar-absorbent materials and a special exterior coating will make the tank difficult to detect at night in the infrared band, while the 'chiseled forms' of the T14 Armata turret will absorb radar illumination."[15] Whether this is true or merely propaganda, the Russians understand the need to mask their tanks and vehicles. A specially developed covering called "Mantiya" is reported to absorb and scatter infrared waves and reduce thermal signatures. So far, this is unproven. In the ongoing Russia–Ukraine War, these claims remain unverified,

Sensors in satellites, aircraft, drones, and ground surveillance systems have created a transparent battlespace. In this photo, a US Air Force E-3 Sentry Airborne Warning and Control System conducts aerial surveillance operations in the Middle East on September 6, 2022. The aircraft provides an accurate, real-time picture of the battlespace. (US Air Force photo by Master Sgt Matthew Plew)

as the T14 has neither been mass-produced nor deployed to combat, as of the time of this writing in May 2023.

Electronic Signature

Masking electronic signatures is a significant challenge as electronic emissions are invisible to the human eye but detectable to the proper sensors. Today, almost all military equipment generates an electronic signature. These include a range of frequencies or wavelengths that cover all forms of electrical emissions, such as the passive operation of any electrical device, radio transmissions, radars, microwaves, and more. Electronic-warfare (EW) systems can also detect radio-frequency (RF) energy without ever emitting a signal. RFID (Radio-frequency identification) tags, used by many units in the US military to keep track of rifles and expensive equipment, can be tracked by both friend and foe, if the sensors are close enough. "When generic RFID chips can be deployed to sense the real world through tricks in the tag, true pervasive sensing can become reality."[16] EW is central to modern combat operations and, in some cases, is more powerful than a kinetic attack. Modern military forces, therefore, are like electric fish swimming in a sea of electrons.

Satellites, aircraft and unmanned aerial systems armed with electronic sensors can "see" emissions from radios, cell phones, mobile computer networks, and even wireless computer printers.[17] They can then send this information to long-range precision fires to target the source. During the first few days of the Second Nagorno-Karabakh War, Azerbaijan used electronic sensors to find Armenian air defenses, EW systems, and command posts. Once these were located, Azerbaijan prioritized its attacks to destroy them. Selecting these high-value targets exposed the rest of Armenia's forces to destruction. Once the Armenian air defenses and EW systems were nullified, Azerbaijan commanded the skies and their loitering munitions and UCAVs hunted artillery, tanks, armored fighting vehicles, wheeled vehicles, and troops, in that order.

In the Russian invasion of Ukraine in February 2022, Russian commanders soon discovered to their dismay that their digital radios did not function as expected. That they were not aware of this before the invasion is telling. Russian contractors installed inferior parts and knock-off microchips into many digital radios.[18] This blatant corruption rendered Russian troops unable to communicate. In desperation, Russian leaders switched to cell phones, using either their own phones or ones looted from Ukrainian civilians. These phones used local Ukrainian cell-phone networks. In some situations, the Ukrainians used these networks against the invaders to spread false information. Other phones used by the Russians were geo-located, eavesdropped on, and targeted by the Ukrainian military, resulting in many Russian casualties and revealing vital combat information such as the location of command posts and ammunition and supply dumps. By February 2023, 12 months after the Russian invasion, the Ukrainians claimed to have located and hit scores of Russian command posts, and many high-ranking Russian officers have been killed. Many of these strikes are confirmed by drone videos. The Russians also reported hitting many Ukrainian command posts. It is clear both sides have learned how to find and destroy high-value targets.

Terrestrial or airborne EW systems can detect military and commercial electronic systems and provide geo-location data. This causes an enormous problem for masking military transmitters, especially for command posts which are, by design, central nodes to collect data across the electromagnetic spectrum. EW systems can even locate secure digital and voice communications. Anything that emits an electromagnetic signature is vulnerable. Minimizing and masking electronic signatures, by fooling the enemy's sensor network with "false positives" and decoys, or hiding within the noise of a city, must become a training priority. Vehicles and command posts that cannot mask will not survive in a transparent battlespace. Masking, therefore, must also become an essential requirement for the design of new tactical systems.

The electromagnetic spectrum is the "high ground" of modern war. Dominating that high ground is vital, but there is a significant gap between what we have and what we need to overcome Russian and Chinese capabilities in the electromagnetic spectrum. If masking electronic and RF signatures is critical to survive and operate in a transparent battlespace, then how can we accomplish this? Perhaps lessons learned by the US Navy's submarine force would prove valuable.

Nicknamed the "Silent Service," submarines survive only when they mask their signature. They have passive means to deceive enemy sensors, such as "silent running" to minimize their acoustic signature; magnetic shielding and degaussing techniques to minimize their magnetic signature; eliminating radio detection by simply not transmitting with radios; using layers of cold water to hide their acoustic signature from active sonar; and reducing the dynamics of a submarine's telltale wake and sound that can be detected under water. Submarines also use active means to disrupt enemy targeting, such as decoys that transmit false signals and EW systems and buoys that jam enemy weapons. Submarines are designed with masking in mind and submarine crews increase their masking ability with well-practiced TTP. Submarine commanders know they must mask their submarine's various signatures from enemy sensors, or they will be located and destroyed.

Land combat offers different challenges, but thinking like a submariner is a logical place to start to improve masking for ground forces. Submarines, for instance, are well known to use silent running to mask sound from sonar detection. In a similar way, ground units could employ electronic silent running by going "dark." The ability to go dark in the electromagnetic spectrum, by turning off all electrical systems, is one option to mask command posts and units. We seldom embrace this option, as it disconnects us from multidomain information, but planning to "go dark" for short, specified periods, would make it difficult for enemy sensor and targeting systems to locate us. If, for instance, a military command post or unit goes dark on a pre-designated schedule prior to an event, such as occupying a position in a city, this would mask the unit from enemy radio-electronic detection. Restricting unnecessary electronic emissions is another means. Limiting the number of systems emitting, and turning off superfluous systems, will reduce the electronic signature. Radio listening silence is the ability to have a communications device tuned to the proper frequency but kept in listening mode. This narrows the chance of being detected by reducing transmissions, as transmitting data requires more power than receiving and is easier to detect by EW systems. Hiding in the noise of a town or a city, a place that usually gives off its own electronic signature, may provide a means to hide a "needle in a stack of needles" by operating in an environment with many "false positive" electromagnetic signatures.

New technologies that make it harder for the enemy to detect our systems are needed. Some of the latest communications devices conceal RF signals. To conceal outgoing signals, tactical communication waveforms rely on transmission security (TRANSEC) techniques to counter eavesdropping, geo-locating, and jamming. TRANSEC secures data transmissions from being infiltrated, exploited, or intercepted by an individual, application, or device. TRANSEC techniques can include frequency hopping, orthogonality hopping, beamforming, burst-transmission, and adaptive processing to develop anti-jam and low-probability-of-detection communications for military operations.

Acoustic and Seismic Spectrum

Any soldier trying to find the Tactical Operations Center (TOC) in the middle of a dark night would take off their helmet and listen for the telltale sound of the diesel-fuel generators that must run 24/7 to provide electricity for the TOC. Human ears are good at detecting sounds, but some machines can discern sounds and determine direction and location better than any human ears. Some of the latest acoustic sensors, mounted on ground and aerial systems, can detect weapons fire and explosions and derive a geo-location. Acoustic sensors can enhance situation awareness in conditions where other sensors might not operate, such as extreme weather, darkness, and beyond line-of-sight. Most ISR drones are called "eyes in the sky," but now, some are also "ears in the sky." Drones equipped with acoustic sensors can locate the fire from artillery systems, with passive sensors that avoid detection by enemy sensor systems, and work beyond line of sight. Ground seismic sensors can also detect vibrations of enemy systems, such as moving tanks, to provide passive, beyond-line-of-sight targeting information.

The latest acoustic vector sensors (AVS) can determine the direction of sound from a wide range of sources, from helicopters to hand weapons. These sensors make it possible to measure the direction of arrival of a sound wave. Only a few millimeters wide, mounting AVS sensors on any military system, including unattended ground sensors, ground vehicles, and unmanned aerial vehicles, is relatively easy. AVS can also detect incoming drones or aircraft, even if they are invisible to radar or do not use RF links. Acoustic-sensor technology is improving and its application to ground and aerial sensors will require masking to avoid detection and targeting. Units will have to reduce the noise signature of their systems and new, more silent systems must be developed and fielded.

Quantum Spectrum

The newest form of target detection and location comes from the weird world of quantum science. This capability is emerging, exotic, and is only being pursued by a few wealthy nations, but it is vital to understand and track as quantum radars are being developed to expose stealth aircraft. These quantum sensors measure variations in acceleration, magnetic fields, gravity, or time. In 2008, a US patent for quantum radar was submitted. The technology is based on the quantum physics of entanglement and is being researched by the US, Canada, Europe, Israel, Russia, and China. On September 3, 2021, the *South China Morning Post*, a mouthpiece

We can learn a lot about masking from submarine warfare, including such tactics as silent running, anechoic tiles (a rubbery product that absorbs sonar waves), the use of decoys, and reducing signals emission. Today, masking from detection of submerged vessels is being challenged by new surveillance and space technologies. In the photo, USS *Santa Fe* (SSN763) joins Royal Australian Navy submarines on training exercises in 2019. (US Navy photo by CPOIS Damian Pawlenko)

for the Chinese Communist Party, reported that People's Liberation Army scientists had made a breakthrough in quantum radar to detect stealth aircraft. "Most radars are known for having a fixed or rotating dish, but this quantum radar looks more like a gun and could accelerate electrons to almost the speed of light," according to Stephen Chen, the author of the article. "The electrons, after going through a winding tube in extremely strong magnetic fields, could produce a vortex of microwaves that barge forward like a tornado."[19]

Whether this is mere propaganda, or an actual breakthrough, is not available in open-source intelligence, but it is clear the Chinese intend to find a means to unmask America's stealth aircraft. For the next few years, this may not be an issue in the current battlespace, as it has limited application for sensing systems on the ground, but its disruptive potential is growing. Someday, quantum sensors will become more common. When this happens, masking in the transparent battlespace will become exponentially more difficult.

A Game of Hide and Seek

Ubiquitous sensors and precision attack is the future of war. New sensors, with sensor-layered networks that mesh sensors from several domains into real-time reconnaissance of the battlespace, make masking essential. While camouflage is a passive means to deny the enemy the optical spectrum, multidomain masking is a full-spectrum approach. Masking, therefore, must be a central theme for every military operation. Modern military forces must either mask or die. As previously stated, masking is the full spectrum, multidomain effort to deceive enemy sensors and disrupt enemy targeting. It is more than passive hiding. Masking is both deceiving and disrupting. It works to deceive enemy sensors through a multidomain effort to depict false positives in the battlespace, to deny the enemy sensor network the ability to coordinate and share information. Masking is central in disrupting the opponent's multidomain targeting system by confusing or jamming the enemy's kill chain and denying the enemy the ability to deliver accurate fires. The concept of masking, therefore, is of supreme importance in war. The old joke that amateurs study tactics and professionals study logistics should be updated. Masters of the modern art of war study tactics, logistics, and masking. Masking, which is much more than security as it involves passive and active efforts by a commander, must be raised to the level of a principle of war.[20]

An example of masking in the ongoing Russia–Ukraine war was the ability of the Ukrainian forces to deceive the Russians during the August–September 2022 Ukrainian counteroffensive. Through a multidomain effort, and with the support of NATO, the Ukrainians convinced the Russians the main effort of their publicized and expected counteroffensive would be in the south, near Kherson, while the main effort was aimed in the north and east of Kharkiv. The Russians moved forces from their

Decoys, dummy vehicles, and fake equipment are a vital part of the art of masking. This inflatable M-47 tank was used by the US Army during the Cold War. During World War II, the Allies fielded mobile, self-contained deception units of inflatable tanks and vehicles to stage multimedia illusions for the Germans. Inflatable tanks and vehicles are also used today by both sides of the ongoing Russia–Ukraine War. (US National Archives photo)

northeastern fronts to the perceived threatened region. As predicted, the Ukrainian attack started in the south on August 29, 2022, and the battle was fierce, with heavy Ukrainian losses. The fury of the Ukrainian attack in the south, and the heavy losses they sustained, convinced the Russians further. Seven days later, with forces masked from Russian observation, the Ukrainians struck in the northeast on September 5, in the Kharkiv–Donetz–Luhansk regions. They overpowered the weaker Russian defenders and liberated over 500 settlements and 12,000 square kilometers of territory in the Kharkiv region. The ability of the Ukrainian forces to mask their preparations for this Kharkiv counteroffensive, by deceiving the Russian sensors and disrupting their targeting network, was a significant accomplishment that aided tactical success. This attack forced the Russians to reassess their war efforts and was a major factor in President Vladimir Putin's order for the "partial mobilization" of manpower in Russia.[21] Manpower matters, but only if these forces can mask on a transparent battlespace.

CHAPTER THREE

The First Strike Advantage

Let your plans be dark and impenetrable as night, and when you move, fall like a thunderbolt.[1]

SUN TZU

He is cold, tired and pissed off. "Damned Russians."

"What?" the other soldier asks.

"Nothing … just thinking out loud." Corporal Jeffrey Stack stomps his feet to keep them warm. He looks out across a snow-filled scene and then glances at his watch. It's 2:40 am and three hours and 20 minutes until they are replaced and rotated back to camp. Three hours and 20 minutes until he can hold a hot cup of coffee in his hand and eat a warm breakfast. Suddenly hungry at that thought, he opens the left pocket of his parka, takes out a cracker saved from his last ration, and munches on it.

The weather is calm with the temperature just below freezing. There is a bright, full moon.

"Well, at least it's not raining," the other man, Private First Class (PFC) Juan Ruiz offers.

Stack smiles, thankful for small blessings. A cold wind from the Baltic Sea has battered them these past few days. It blew so strong it disrupted drone and aircraft operations. Rain, followed by sleet, and then snow, reduced visibility to a few meters. Not even helicopters could fly in those winds, but now, all is tranquil. Stack can see past the open field to the next tree line, a kilometer away. Three more hours, then back to camp, warmth and hot chow.

His mind drifts to different times. The scene is so peaceful it reminds him of winters back home in Wyoming. He wishes he was back home right now, sitting at the dinner table with his mother and father, watching football on television and telling Army stories to his younger brother, or relaxing by the fireplace, discussing the latest news with his uncle. At home he would be warm, comfortable, and safe. It seems hard to believe that only a month ago he was there, enjoying a 10-day leave. Now, he and many other American soldiers are on guard in Lithuania.

Ruiz stands by Stack's side, shivering in the cold. The two men occupy a deep foxhole observation post (OP) in a snow-covered forest of conifer trees 25 kilometers from the Russian border. The open ground in front of them is flat and covered with a foot of snow. Their unit, Lightning Troop, 3rd Squadron, 2nd Cavalry Regiment, occupies an assembly area in the forest two kilometers from his OP, near the town of Pabrade, Lithuania. Watching from their foxhole at the edge of the trees, all is quiet.

Stack reflects on the events of the past few weeks. Maybe this will all blow over and nothing will happen, he thinks. International tensions are high. It's been nearly a year since the cease-fire ended the Russia–Ukraine War. Belarus is no longer independent, but has merged with Russia, and is full of Russian combat units. The Russians have mobilized new manpower and rebuilt forces they lost in Ukraine. Rearmed with new, Chinese-made tanks, artillery and drones, the Russians are saber rattling. They want a wider corridor from Russia–Belarus to their enclave at Kaliningrad, calling it their historic Russian land, and they demand the removal of "provocative North Atlantic Treaty Organization (NATO) forces from the Baltic nations." The area the Russians want is called the Suwalki Corridor and NATO has drawn a line in the snow and said "No, you can't have it." The Russian leader then threatened that if NATO forces didn't withdraw from the Baltic countries in 48 hours, it would mean war. That deadline passed yesterday, when the weather was at its worst.

Stack and Ruiz stand in their OP with weapons loaded, waiting for something to happen and hoping nothing does. They are the forward edge of the thin green line in the snow to deter the Russians.

"Another miserable night," PFC Ruiz announces. "I hate snow … I hate this is on-alert, off-alert shit … hell, I'm even beginning to hate you."

"Hell, I'm the best friend you've ever had."

"You think the Russians will do anything?"

"Don't worry," Stack laughed. He enjoyed their banter in times like this. It helped him forget how miserable they were.

There was a long pause. Stack thought he heard something, then dismissed it.

"Well? You too cold to talk?" Ruiz prodded.

"I don't think anything will happen. This isn't like Ukraine. The Russians know we're here and they'll back down, you watch. Those bastards don't have the balls to attack us."

"Yeah, I heard the company commander talking about it. So, we're a tripwire, huh?"

"Yeah … NATO told Moscow not to cross the border or else. That's why we're here. The Russians are bluffing. An attack would mean all-out war, maybe even nuclear war. Think about it."

"Cheery thought," Ruiz answers. "What if they're not bluffing?"

"Hmph … Not likely. They got their nose bloodied in Ukraine."

"Tripwire … yeah. That's what we are."

A buzzing sound suddenly breaks the quiet of the night. The two soldiers stare into the darkness. Stack flips down the night-vision goggles attached to his helmet.

"Can you see it?" Ruiz asks as he scanned the sky.

"No, but I sure do hear it! It must be nearby."

Then, 10 feet in front of them, a black quadcopter drone appears, hovering over the snow-filled field. After a few seconds it buzzes away, flies skyward, but then returns to hover six feet off the ground in front of the Americans. Stack looks at the drone through his night-vision goggles. Man and machine stand frozen, locked in an eerie moment, staring at each other.

"I see it! Should I fire?" Ruiz shouts, pointing to the drone.

"Hold your fire. Maybe it's one of ours. Let me confirm first."

The drone shoots upward and flies over the trees.

Stack keys his handheld radio and asks his platoon leader for advice. Minutes pass.

Then, like a mighty hammer from the God of War, the air fills with the sound of incoming rockets and artillery shells. The scream of the barrage is deafening. Dozens of sub-munitions explode in a mighty roar among the trees behind Stack and Ruiz. Trees fall and wooden splinters scream through the air. Slammed into the ground from the concussions of the shells, the two Americans die when thermobaric shells explode overhead and ignite into a fireball that burns everything. The thin green line in the snow is about to be crossed.

***

This story is fiction, but no matter what happens in the ongoing Russia–Ukraine War, a similar scenario could happen to American forces tomorrow. US troops are deployed in the Baltic nations, eastern Europe, and in about eighty countries worldwide. They are often used as a "tripwire," a presence that acts as a deterrence to war, such as in the Republic of Korea. Unfortunately, a tripwire only works if the opponent fears tripping the wire. What if enemies no longer fear retaliation? What if our forces, deployed in small numbers in the Baltic countries, Eastern Europe, or the Pacific, are located and targeted by our enemies? Many of these units occupy fixed and known locations. Some units rotate in and out of the same areas, like US forces in the Republic of Korea. Our adversaries know where they are. We assume our troops deter the enemy from going to war. What if the other side sees our troops, not as a tripwire, but as an easy target? Are our forces training to survive a surprise first strike? If not, why not?

Dictatorships understand the value of striking with surprise. It is a rule, not an exception. Numerous examples exist, such as the German invasions of Denmark, Norway, France, Belgium, and Holland in the spring of 1940; the German invasion of Yugoslavia, Greece, and Russia in 1941; the Japanese attacks on Pearl Harbor and across the Pacific on December 7, 1941; the invasion of the Republic of Korea by Communist North Korea on June 26, 1950; the Chinese surprise attack against US and United Nations forces in the Korean War on Thanksgiving Day, November 25, 1950; Saddam Hussein's invasion of Kuwait on August 2, 1990; the surprise Russian attack and annexation of Crimea on February 20, 2014; the Russian entry into "main force" operations with the attack on Ukrainian forces near Zelenopillia on July 11, 2014; and the Russian "Special Military Operation" attack on Ukraine on February 24, 2022. These are some of the most noteworthy "first strikes" by totalitarian powers in the 20th and 21st centuries. The idea is to strike first with surprise against an unsuspecting enemy, strike hard and stun the opponent with overwhelming force. Today, with weapons of much greater range and precision, the first-strike advantage is a powerful weapon in the arsenal of war strategies. It can create battleshock. Our adversaries embrace this idea. Deploying forces within range of an enemy first strike without preparation to ensure their survivability is pure folly. With the US military placed in vulnerable positions around the world, will its forces survive the enemy's first strike?

Multiple Launch Rocket Systems (MLRS) deliver saturation fire—intense and rapid bombardment. The MLRS is an armored, self-propelled, long-range precision-fire weapon which is used to engage enemy targets over 175 miles (282 kilometers) away, and move to different locations rapidly to avoid enemy counter fire. The rockets can be fired individually or in ripples of 2 to 12. Accuracy is maintained in all firing modes because the computer re-aims the launcher between rounds. In this photo, US Army MLRS of the 147th Field Artillery launch rockets during a training exercise on August 2, 2015, at Camp Guernsey, Wyoming. (US Army National Guard photo by Senior Airman Duane Duimstra)

Russia has been perfecting the ability to tie sensors to long-range shooters to destroy high-value targets in near real-time for over 50 years, but never had the technology to make it work. Now it does. "The Reconnaissance Strike Complex" [разведивательно-ударный комплех, RYK] was designed for the coordinated employment of high-precision, long-range weapons linked to real-time intelligence data and precise targeting provided to a fused intelligence and fire-direction center."[2] The reconnaissance–strike and conventional artillery capabilities of America's peer enemies present a terrifying equation: everything on the battlefield will be located, targeted, and hit. With multidomain sensors capable of creating a transparent battlespace, and long-range precision fires able to reach anywhere, there is no sanctuary in the modern battlespace. In any future conflict with Russia or China, the side that strikes first has a marked advantage. If future war can be analogous to a chess game, but now one that is played in five dimensions, then the Russians and Chinese are playing the white pieces and thus have the first move, the benefit of launching a first strike. In such a case, the survival of troops and vehicles in the extended battle area, which extends across the entire battlespace, will be in jeopardy. Unless we can dig in quickly and deeply enough to protect our forces from this intense firepower, our only remaining option is to mask.

Attacking first, especially when the opponent is unaware or unprepared, can produce significant advantages and is a time-tested tactic of warfare. Today's advanced sensor networks and long-range precision fires make a non-nuclear first strike a tantalizing option for any power wishing to turn the balance of power against NATO and/or the United States. As the US looks to deter further Russian designs on Europe, the advance of the Chinese on Taiwan and the South China Sea area, or the actions of an unpredictable North Korea, the advantage offered by an enemy first strike becomes our nightmare scenario. Russia invaded Ukraine in 2022 for several reasons, but one was to secure a land bridge to Russian-occupied Crimea. Russia also occupies the enclave of Kaliningrad. A land bridge between Russia's ally Belarus and Kaliningrad, called the Suwalki Corridor, has been a long-term Russian goal. NATO only positions battalion- and brigade-size combat forces in the Baltic countries, and these forces are on short, rotational deployments. For NATO forces in Latvia, Lithuania, Estonia, Poland, and Romania, for instance, the Russians know where they are and could launch a devastating first strike against them with little to no warning. It remains conjecture as to Russia's willingness to risk this, but NATO forces cannot count on Russian intentions.

The same is true in the Pacific. Whether Xi Jinping intends to order the Chinese Communist Party's People's Liberation Army (PLA) to attack Taiwan is unknown but, if he does, it will start with a massive, unannounced, surprise attack. The PLA is not sure if the US will rush to the aid of Taiwan. We have excelled at strategic ambiguity in this case, but the PLA would not make an attack on Taiwan without taking our reactions into consideration. Ignoring the Americans is not an option.

Just as the US Navy's fleet was in the way of Imperial Japanese ambitions in World War II, the ability of the US to intervene with multidomain forces in an attempt by the PLA to take Taiwan would be the prime consideration of PLA planners. War games conducted by the Center for Strategic and International Studies (CSIS) in January 2023 of a Chinese invasion in 2026 predicted the PLA would have a hard time taking Taiwan. If the US intervened, these games predicted the Chinese would lose, but with heavy American casualties. "The United States and its allies lost dozens of ships, hundreds of aircraft, and tens of thousands of service members. Taiwan saw its economy devastated. Further, the high losses damaged the US global position for many years. US power projection would be degraded for years."[3] In every one of the 24 scenarios run in these war games, the Chinese struck first with rockets, drones, and aircraft and their first strike jeopardized the survivability of American aircraft carriers and bases in the Pacific region.

The PLA will hit these "most dangerous" targets first. "China has one of the largest and most diverse missile arsenals of any country in the world. **And that arsenal is growing**."[4] In spite of the "optimistic" predictions of the CSIS war games, if the US is unprepared when the Chinese move, a PLA first strike on American forces in the Pacific will make the Japanese attack on Pearl Harbor look like child's play.

With today's sensors and long-range precision weaponry, a well-executed first strike is the surest means to begin a conflict with a tactical, operational, and strategic advantage. The latest publication of the US Army's capstone manual, *Field Manual 3-0* (October 2022) states: "Forward-based Army forces in range of adversary fires require significant hardening for survivability against enemy ballistic missiles, aircraft, naval fires, and cyberspace attacks. Forward-based Army forces can defend critical joint infrastructure when properly positioned and prepared."[5] To ignore this possibility of an enemy first strike, in Europe or in the Pacific theater, is foolhardy, dangerous, and criminal. Opponents, bent on military action, will almost always have a first move against US forces in the opening stage of a war. Even if they cannot match the US and its allies on a global level, they will develop operational and tactical niche capabilities to strike first with devastating effects. Surviving the first strike requires a different mindset than what we have today.

All US military units must train for this possibility. For instance, US Army units rotate to several combat training centers (CTC) to conduct training for war. These rotations typically involve a brigade-size unit with supporting units and sometimes a Division Tactical Operations Center from the brigade's parent division. The method used to move to the CTC follows a legacy, *Desert Storm*-like model. The unit moves by rail to the CTC, conducts an off-loading and maintenance period, prepares for combat operations, and then moves into the training area to conduct operations. This method is conventional, reasonable, and flawed. No future opponent will be as insane as to allow the US to move equipment and forces, a "mountain of iron," into position to launch an attack as the US did in Operations *Desert Shield/Storm*

Conventional Strike Capabilities

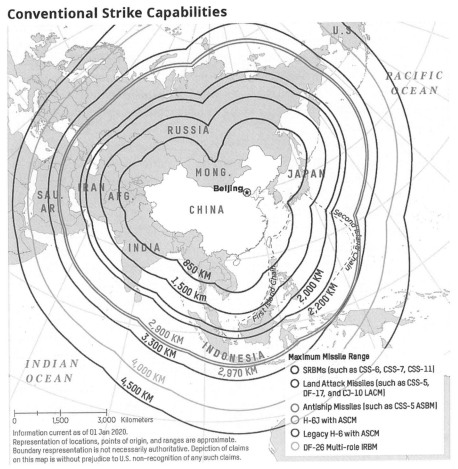

This image shows the range of Chinese missiles launched from China. US forces in the Pacific are in reach of a Chinese first strike. Chinese missiles can reach US bases in Japan, the Republic of Korea, Guam, the Philippines, and Hawaii. (Secretary of Defense: Military and Security Developments Involving the People's Republic of China 2020 Annual Report to Congress, 58)

and *Iraqi Freedom*. To train for future warfare, the US Army should start deploying units as they arrive, directly from the railhead to assembly areas where they must mask to avoid the enemy's first strike. The unit should not have the luxury of an "administrative" period to unload and get ready for operations. In future wars, if the enemy has learned anything from the American way of war, US forces must learn to mask as soon as they enter the contested battlespace. At the CTCs, we must train as we will fight.

The importance of being prepared to fight on the first day of conflict was demonstrated during the Second Nagorno-Karabakh War (2020). Azerbaijan struck first against unprepared Armenian positions. Employing newly purchased high-tech unmanned aerial vehicles (UAVs), loitering munitions, and conventional artillery and rockets, the Azerbaijanis destroyed the Armenian air-defense and command-and-control network within the first week of the war. Armenian air defense, made up of older Russian-made systems, could not stop the Azerbaijani UAV and loitering-munition attacks. Even when the Armenian air-defense systems were operating, the aerial top-attack weapons penetrated the airspace and knocked out the defenders.

Azerbaijan used a wide variety of UAVs during the war. The most effective and notorious were the Turkish-made Bayraktar TB2 (Turkish defense company Baykar) unmanned combat aerial vehicles (UCAV), the Israel Aerospace Industries (IAI)-made Harop, and the fully autonomous Israeli-made Skystriker (Elbit Systems). The electro-optical sensors on the most sophisticated UAVs and loitering munitions used by Azerbaijan were state-of-the-art infrared and low-light high-definition television cameras that delivered secure tactical video reconnaissance, surveillance, and targeting data. During the war, the TB2 operated as an attack platform and as the "eye in the sky" for Azerbaijani forces to identify and designate targets for other UAVs, UCAVs, loitering munitions, artillery, rockets, and smart anti-tank guided missiles (ATGM), such as the Israeli-made Spike ATGM system. While the TB2 UCAV identified Armenian forces in the designated strike zone, loitering munitions circled overhead, verified their targets, and then dived into their victims to detonate their 23-kilogram (51-pound) warheads in a kamikaze-like attack. Prior to launch, the Harops are programmed to fly, autonomously, to a pre-defined strike zone. Once there, they loiter and the human operator can select one loitering munition for target search and attack, while others orbit the strike zone. According to IAI, "The operator directs the selected LM (Loitering Munition) to the target area and uses the video image to select a target, and to attack it. The Harop tracks the target and then dives on it, detonating the warhead upon impact. If required, the attack can be aborted, and the operator can re-attack with the same LM."[6] A vital component is the camera system that allows the operator to see the battlespace in real-time and direct the loitering munition to attack a specific target.

Unmanned Combat Aerial Vehicles and Loitering Munitions

During a single strike in October 2020, Azerbaijani TB2 UCAVs, working in a pack with Harop and Orbiter loitering munitions, annihilated eight Armenian howitzers dug into protected firing positions. Unmanned aerial systems (UAS),[7] and especially UCAVs like the TB2, and loitering munitions furnished Azerbaijan with a decisive advantage over its Armenian opponents. Azerbaijan repeated these attacks against artillery, armor, command-and-control facilities, and troop positions throughout the Second Nagorno-Karabakh War.

In March 2022, as eight Russian 122-mm D-30 howitzers blasted away at a Ukrainian town in the Kherson region, a Ukrainian TB2 flew into striking range. The Russian howitzers were deployed in a line in an open field. Unseen by the Russian gunners, and unidentified by air-defense units protecting them, the TB2 laser-designated a Russian howitzer and dispatched a missile. Miles away, safe from enemy fire, a Ukrainian drone operator "flew" the TB2 to a higher altitude to get a better view of the strike. Watching his screen, the TB2's sensors streamed real-time video of the missile's progress to the controller. Within seconds, the Russian howitzer erupted in flames and the Ukrainian drone pilot cheered. He then continued to fly the TB2 to fire at another howitzer. After expending all its missiles, the TB2 operator then designated the remaining Russian guns for artillery fire, forcing the Russian artillery battery out of action.[8]

A New Form of Striking Power

UCAVs are multipurpose craft and can perform reconnaissance, designate targets with on-board lasers, and act as robotic missile and bomb launch platforms. An example of the latest generation UCAV is the US Predator MQ-20 (formerly Predator C) Avenger. The MQ-20 uses a jet engine and is one of the most capable UCAVs in the world; it is also one of the most expensive. With an estimated unit cost between US$70–150 million (exact costs are classified), the Avenger can fly at 460mph, at a maximum altitude of 50,000 feet, remain airborne for up to 18 hours, and has points for six Hellfire missiles or other bombs. Avengers typify the weapon of choice for penetrating and defeating enemy anti-access and area denial zones, a prerequisite to attaining air dominance.

In comparison, the Turkish TB2 offers a less expensive choice for reconnaissance and precision attack. Each propeller-driven TB2 costs US$3–10 million, a bargain when compared to an Avenger. The TB2 is a medium-altitude, long-endurance UCAV that is slow and not very stealthy, but it continues to make headlines in the ongoing Russia–Ukraine War. Ukraine purchased 20–30 TB2s prior to the invasion and employed them against the Russian Army with success. The current version of the TB2 used by the Ukrainian Army employs a direct line-of-sight radio link. The Russian Army's advanced electronic-warfare (EW) systems should have been able to jam this linkage, and Russian gun and rocket air-defense systems should have swept the sky clear of TB2s, but Russian counter-unmanned aerial system (CUAS) defense in the first three months of the war was haphazard.

This implies that Russian CUAS and EW jamming were not optimized to defend against this threat in the first months of the war. Several reports credit TB2s with destroying the Pantsir-S1 air-defense system, a weapon designed to defeat aircraft, helicopters, precision munitions, cruise missiles, and UAVs. The system's phased-array radar has difficulty identifying slow-moving threats like the TB2. The TB2 may also have played a role in defending against Russian naval forces as multiple, independent reports purport that the sinking of the Russian guided-missile cruiser *Moskva* involved a missile and drone attack that included the use of at least one Ukrainian TB2.

Loitering munitions provide another means to attack deep behind enemy lines. Loitering munitions are "smart missiles," equipped with explosive warheads and guidance suites that loiter over an area, search for targets, and then strike. The US firm AeroVironment refers to their loitering munitions as "loitering missiles." A capable and cost-effective means for precision attack, loitering munitions are becoming a key weapon system in the modern, multidomain battlespace. In any military engagement, timing is vital, and loitering munitions deliver a "loiter, sense and strike" ability that can reduce targeting time from minutes to seconds. Unlike traditional aircraft, most loitering munitions require little or no runway; most launch from rails or can take off vertically. Some loitering munitions, such as the IAI Harop, launch from truck-mounted firing boxes, much like a multiple rocket launcher. The Harop can loiter over a target area for six hours (land version), or nine hours (sea version), and carry inside the airframe a high-explosive fragmentation warhead comparable to a standard NATO 155-mm howitzer projectile. If the Harop

The Marine Corps is acquiring more small UAS vertical take-off and landing (VTOL) systems. In this photo, US Marine Corps Sgt Danielle Grimshaw employs a US-made R80D Skyraider in support of engineer reconnaissance, during Exercise *Archipelago Endeavor 22* at Berga Naval Base, Sweden, on September 20, 2022. (US Marine Corps photo)

cannot identify a suitable target, it shifts to a subsequent priority or, if no target is found, it will return to a designated landing point on its own for retrieval and reuse. There are a wide variety of loitering munitions, from short-range versions such as the American-made Switchblade, to the longer-range Harop. Because of their low cost compared to the price of their targets, loitering munitions are a cost-effective means of delivering persistent, precision strikes.

Adapting to the New Reality of Drone Warfare

UCAVs such as the TB2 are not invincible. Over time, the Russians learned, adapted, and now employ an integrated air-defense system and EW network in depth that make it difficult for TB2s and other UCAVs to operate in the contested battlespace. Many were shot down or jammed by Russian EW systems. To address this problem, Baykar has created an updated TB2S version that is controlled by a satellite link, similar to the more expensive US-made Predators. The TB2S will be much harder to defeat, by breaking the signal to the operator, because of this connection. It will cost more, but it can operate against advanced EW and jamming systems. In addition, Baykar is developing a TB3 variant. This next-generation system will take off and land on shorter runways, or aircraft carriers as it has folding wings for placement on ships, and can carry six missiles, two more than the TB2. Like its predecessor, the TB3 will conduct reconnaissance, surveillance, intelligence, and attack missions. According to Baykar, the TB3 will have a satellite control system that provides both line-of-sight and beyond-line-of-sight communications capabilities mimicking the more expensive US and NATO UCAVs. This will permit the control station for the TB3 operator to be located anywhere on earth as long as it maintains a satellite connection. Baykar reported the TB3 could be available as soon as 2023. Baykar also purchased land in July 2022 to build a drone factory in Ukraine.[9] Although Ukraine does not have the TB2S or TB3 at the time of writing, it may get an upgrade in the future.

Humans control most of the current generation of UCAVs and loitering munitions via a data-transmission link, but soon automated and artificial-intelligence systems will replace the need for a human operator. The role of the operator is shifting from controlling the system to commanding the system. Commanding the system will entail activating it to fly and strike targets inside a designated strike zone according to preset targeting priorities. The system will automatically execute its pre-programmed mission, but a human may still be in the loop to intervene and call off, or switch, the point of attack. This shift from "control" to "command" will become possible as drone manufacturers create smarter systems. The Elbit Systems Skystriker, for instance, is a fully autonomous loitering munition that can locate, acquire, and strike operator-designated targets on its own. If desired, systems like these could hunt with minimal human interaction inside a designated battlespace. Western nations are keen to keep a human in the loop, but Russia and China may not be so reluctant to remove this aspect.

Unmanned combat aerial vehicles are changing the face of war. This photo shows a US MQ-9 Reaper taking off from an Air Force Base in Nevada on December 17, 2019. The Reaper is a high-level, remotely piloted weapons platform capable of instant action and precise engagement. The Reaper was introduced in 2007 and has been battle tested in combat around the world. The US Air Force is looking for newer systems. Northrop Grumman has proposed a plan to replace the MQ-9 Reaper with stealthy autonomous drones, which could be used in highly contested environments such as the Indo-Pacific region. (US Air Force photo by Senior Airman Haley Stevens)

Dominating the battlespace with persistent and pervasive precision fires, especially for decisive engagements or targets, will become a battle-winning tactic. As unmanned systems become faster, smarter, and more lethal, loitering munitions and UCAVs will deliver the means to accomplish this objective. UCAVs and loitering munitions are force multipliers that will transform the methods of warfare. It is realistic to envision that, in the near term, every mortar platoon, reconnaissance, and maneuver organization will have loitering-munition units, and artillery organizations will contain loitering-munition batteries and UCAV squadrons. This does not mean the tank is dead, as it provides the means for offensive maneuver, something drones do not, but tanks will only survive and win if they adopt capabilities that adapt to the new battlespace. Those who recognize this, and prepare, will gain a tremendous tactical and strategic advantage.

Strike First and Strike Hard

Today, ubiquitous sensors can detect almost everything in the battlespace, and long-range precision fires and drones can capitalize on these capabilities to destroy high-value targets (HVT). The first-strike advantage is the ability of an attacker to paralyze an enemy in the first hours and days of a war. A surprise first strike that destroys the most critical targets, if executed with overwhelming force, can be decisive. The Azerbaijanis achieved this in the first weeks of the Second Nagorno-Karabakh War. The Russians did not maximize the first-strike advantage when they invaded Ukraine. On March 10, 2022, Russian Defense Ministry spokesperson Igor Konashenkov said Russia hit 2,911 Ukrainian military facilities.[10] Even if this number was accurate, and not propaganda, it did not break Ukrainian defenses or their will to fight. Despite hundreds of Russian artillery, missile, and air strikes in the war's first week, Russian long-range precision-fire attacks were inadequate for the scale and depth of the battlespace. Russia hit key targets, such as Ukraine's internet service and communications capabilities, but in two weeks the internet was back in operation thanks to American entrepreneur Elon Musk who provided Ukraine with his Starlink internet service. The Russians failed to eliminate other HVTs, such as capturing or killing President Zelensky and destroying key Ukrainian government

The Orlan-10 is an unmanned aerial vehicle developed by the Special Technology Center in Saint Petersburg for the Russian Armed Forces. It is primarily used for intelligence, surveillance, and reconnaissance, and artillery spotting. In this photo, a Ukrainian soldier (face concealed for security reasons) holds an Orlan-10 his team was able to down with electronic jamming in May 2022. According to *Reuters*, the Orlan-10 has been used in the conflict between Russia and Ukraine, directing up to 20,000 artillery shells that have killed up to 100 soldiers per day. (Ukrainian Ministry of Defense photo)

facilities and headquarters. Instead of decapitating the enemy and forcing an immediate surrender, the Russians only stiffened Ukraine's resolve and resistance. You can be sure the Chinese have studied this "first-strike failure" and will plan for an overwhelming first strike to capture Taiwan.

An enemy first strike is a serious threat to the US and its allies. China or Russia may seize the opportunity to cause a crisis, watch opposing forces move into nearby countries, locate them, and then launch a surprise attack. Russia's surprise first strike against Ukraine on February 24, 2022, was not hard enough. If a resurgent Russia gets a second chance to launch a first strike, they will have learned they must go "all in" to make it decisive. China has also learned this lesson.

If the battlespace is transparent, how is a first strike possible? Surely, the enemy will be seen positioning in time to prepare for a first strike. If the past is prologue, we should learn our best intelligence and surveillance will not warn us in time. In December 1941, we had more information than we could consume. "If our intelligence systems and all our other channels of information failed to produce an accurate image of Japanese intentions and capabilities," Roberta Wohlstetter wrote in her 1962 book *Pearl Harbor: Warning and Decision*, "it was not for want of the

The Bayraktar TB2 was a critical weapon in Azerbaijan's victory in the Second Nagorno-Karabakh War. According to the open-source intelligence Oryx website, in 44 days of conflict, a handful of Azerbaijani TB2s essentially broke the back of the Armenian military, destroying a confirmed total of 549 ground targets including 126 armored fighting vehicles (including 90 T72 tanks), 147 artillery pieces, 60 multiple rocket launchers, 22 surface-to-air missile systems, six radar systems, and 186 other vehicles. (Baykar Technology)

relevant materials. Never before have we had so complete an intelligence picture of the enemy ..."[11] US decision makers:

> ... did not have the complete list of targets (estimated to be the objectives of an evidently imminent seaborne attack), since none of the last-minute estimates included Pearl Harbor. They did not know the exact hour and date for opening the attack. They did not have an accurate knowledge of Japanese capabilities or of Japanese ability to accept very high risks ... If we could enumerate accurately the British and Dutch targets ... [of] a Japanese attack ... either on November 30 or December 7, why were we not expecting a specific danger to ourselves?[12]

Technology has changed significantly since 1941, but human decision making is roughly the same. Terrorists surprised the US on September 11, 2001, and the US had far better intelligence-collection capabilities than in 1941. Tomorrow, if our adversaries decide to attack us, we will be hit with an overwhelming first strike by precision missiles, artillery, and drones. Are we prepared?

Attacking targets from the above is the preferred method of engagement in modern war. In this photo, a Northrup Grumman & Shield AI V-BAT, vertical take-off and landing Unmanned Aerial System (UAS), takes off. No launch system is required. The V-BAT has been selected by the US Army to replace the long-serving RQ-7B Shadow UAS. (US Army photo)

CHAPTER FOUR

Top Attack

Untutored courage is useless in the face of intelligent, precision weapons.[1]
THE AUTHOR PARAPHRASING A QUOTE ATTRIBUTED TO GEN. GEORGE S. PATTON, JR.

"Contact!" A Ukrainian drone operator named Kitko shouts.

Kitko pilots a Bayraktar TB2 unmanned combat aerial vehicle (UCAV) from inside a mobile control station near the city of Malyn, about ninety-six kilometers (sixty miles) northwest of Kyiv. Russian vehicles are stalled along the road to Kyiv, each vehicle only spaced a few meters apart. It looks like a big city traffic jam during rush hour.

Kitko smiles. "We have them!"

It is February 27, 2022, and Ukraine is at war, fighting for its survival. Russia invaded only a few days ago. The Russian first strike hammered the Ukrainian Army, and at first the advancing Russian columns seemed unstoppable. In these beginning days, everything is confusion and chaos. The winter sky is gray, snow covers the ground, and the green Russian vehicles are easy to see using the TB2's camera-sensor.

Kitko's Turkish-made TB2 is armed with four MAM-L missiles. MAM stands for "Mini Akıllı Mühimmat," which in Turkish means "smart micro munition." The MAM-L's range is 15 kilometers (9 miles). The missile's powerful tandem warhead—a combined warhead comprised of two projectiles that detonate a few milliseconds apart after contact with the target in order to punch through thick armor—is enough to destroy or disable a tank. Seeing the target with the TB2's sensor suite and then attacking the victim right away with on-board missiles quickens the kill chain. The time between sensing and striking can be seconds, not minutes. One of the TB2's main advantages is this sense-and-strike capability.

From 7 kilometers (4 miles) away, the TB2's gyro-stabilized infrared camera focuses on a Russian Buk M2E Air Defense System (NATO reporting name SA-17 or Grizzly) halted along the road. The Buk is an impressive air-defense weapon and, if the system is on, and the crew is vigilant, it can detect and take down the TB2. If the Buk is tied into other air-defense systems, such as the Pantsir,[2] in an integrated air-defense network, they may jam Kitko's signal and disable his TB2.

Kitko takes a deep breath. A Pantsir may be positioned nearby; it is hard to be sure. He knows he is the hunter, but his prey is a wolf, among a pack of wolves, and wolves are dangerous beasts. He knows he must stop the Russians. If they can destroy the Russian air-defense systems, the TB2s will be able to roam the skies and decimate the invaders.

"I see a high value target. There are two, no three… confirmed… Buks … they're on the road." Kitko guides the TB2 over a clump of forest and then soars higher to have a look at the stalled column. The TB2's high-magnification electro-optical and infrared targeting and designating sensor provides a stabilized, close-up image from many kilometers away.

The atmosphere turns electric with excitement. Two other Ukrainians occupy the mobile control station. A second operator, a sergeant named Demchak, sits at a console to Kitko's right. Demchak pilots a second TB2. Since there are only two TB2s available, the third station is unoccupied. A captain stands behind the two drone pilots, watching intently as he smokes a cigarette, but says nothing.

As he pilots the unmanned aircraft, Kitko considers what a godsend the drone is to Ukraine. The Russian's first strike destroyed many of his army's artillery and rocket units, leaving little striking power for Ukraine to hold back the invaders. Sadly, Ukraine only has a few of these drones, but what they lack in numbers they make up for in skill and daring. It takes a well-trained pilot to fly the TB2 to maximize its killing potential, and Kitko is one of the best.

Kitko's greatest concern is to keep his TB2 "alive" and make every missile count. The Ukrainians have fewer than 20 operational TB2s and they will soon be out of missiles. The TB2 can carry four missiles, but the hardpoints on the TB2's wings can only mount the specially designed MAM missiles. Without a deep stockpile of these missiles, his TB2 will soon become a scout, not a hunter.

He observes his control screen and sees two Russian soldiers standing outside the target vehicle, smoking cigarettes. He studies the target and sees that the Buk's radar isn't moving. He smiles. "Do you see them?"

"*Tak*," Demchak answers with a wide grin. "They're just sitting there on the road. My bird is moving up from the south, about two kilometers behind your right wing. Now is the time to strike."

"Their radars are not deployed," Kitko offers. He observes several 4×4 armored trucks, most likely GAZ 2330 Tigers along with three Buk air-defense systems. Kitko knows from experience a Buk missile battery consists of two TELAR vehicles (Transporter Erector Launcher and Radar, with four missiles apiece) and one TEL (Transporter Erector Launcher, with six missiles) for a full complement of 14 missiles. Most likely, this was a Russian air-defense battery trying to move forward to provide air-defense coverage. Kitko knows if the Russians haven't deployed their radars, they can't fire their missiles.

"I'm in position to fire," Demchak reports.

"The Buks are the prize. Let's hit 'em before they can set up." Kitko replies as he soars the TB2 skyward and prepares to fire a MAM-L at his target.

"I'll call in artillery fire," the captain behind them announces. "We have a 152-mm howitzer in range."

"I have the trail vehicle in my camera view," Demchak reports. "They still don't see us coming. Arming missiles. Ready to fire when you are."

"Arming missiles," Kitko announces. "I'll hit the first vehicle. You hit the last one. Then we will work our way down the line as the artillery flattens them."

"Of course, and I'll video while you strike," Demchak added. "Let's give these bastards some Turkish brew this morning."

"Firing now! *Slava Ukraini*!" The real-time video on his control screen gives Kitko a clear view of his target as he presses the trigger. At nearly the same time, Demchak fires. The missiles lock onto the targets and fly true. In seconds, two Buks explode in brilliant red, orange and yellow fireballs.[3]

***

The narrative above is fictional, but the action was real and occurred a few days after the Russian invasion of Ukraine.[4] Although Ukraine only had a few TB2s, they had a dramatic impact on the tactical situation in the first months of the war. The use of top-attack drones was foreshadowed in the conflict in the Caucasus two years earlier when Azerbaijan and Armenia clashed over Nagorno-Karabakh. The terrain there is mountainous and offers advantages to the defender, but Azerbaijan used technology and cross-domain maneuver to win its rapid and decisive victory. The Azerbaijanis invested a reported US$24+ billion to upgrade their forces prior to the war, purchasing the latest Turkish and Israeli unmanned aerial vehicles (UAV) and loitering munitions. The effective use by Azerbaijan of top-attack weapons devastated and demoralized the Armenians. These systems did not win the war by themselves, the Armenian air defenses were flawed, and UAVs have been over-hyped in many press accounts of the war, but the impact of the high-definition, full-motion video provided by top-attack systems categorized this conflict as "a milestone in military affairs."[5]

In Ukraine, information from unmanned aerial systems (UAS), satellite images, and other sensors, depict the exact location of Russian movements. Unable to mask their forces, the Russians became extremely vulnerable to this accelerated sense-and-strike kill chain. If the Ukrainians had invested in more UCAVs and loitering munitions, they may have caused a heavier toll on the invaders early in the war. Recognizing this, the US offered shipments of AeroVironment's Switchblade to Ukraine on March 17, 2022. The Switchblade loitering munition saw combat in Afghanistan with US Special Forces.

Unmanned, top-attack systems play a critical role in modern warfare. UCAVs and loitering munitions provide a relatively inexpensive, sense-and-strike capability

Top-attack munitions are the preferred method of attacking armored vehicles in modern war. US soldiers assigned to the 2nd Platoon, Alpha Company, 1/163rd Combined Arms Battalion, launch a Javelin shoulder-fired anti-tank missile during a live-fire exercise in Syria on March 25, 2022. Top-attack munitions, such as the Javelin, were used by the Ukrainian Army to defeat Russian armor in the opening months of the Russia–Ukraine War in 2022. (US Army photo by Spc William Gore)

to accelerate the kill chain. These unmanned aerial sense-and-strike systems are hard to counter. They came of age during the Second Nagorno-Karabakh War and have improved since then, getting smarter, more precise, faster, and stealthier. In the Israel–Hamas War (2021), Israel Defense Forces (IDF) used state-of-the-art drones to dominate the skies and provide persistent intelligence, surveillance, and reconnaissance (ISR) information over Gaza. Providing real-time ISR, the IDF knew where their opponent was and followed their movements. The IDF demonstrated the fusion of the transparent battlespace with top-attack precision munitions to defeat Hamas.

In February 2022, few military analysts expected slow-flying drones to be a problem against Russia's advanced air-defense systems. Ukrainian UCAVs, however, inflicted devastating attacks on the Russian Army in the opening months of the war. The TB2 became famous in the beginning weeks of the Russian invasion as it struck stalled Russian vehicle columns lining the roads to Ukraine's capital, Kyiv. The TB2 also scored a dramatic victory when it played a part in a combined attack, that involved at least one TB2 and several Neptune missiles, to sink the guided-missile cruiser *Moskva* on April 14, 2022.[6] *Moskva* was the flagship of the Russian Baltic Fleet and its sinking was a major military and propaganda victory for Ukraine. High-end, air-defense systems built to detect fast-moving aircraft and missiles appear to have a difficult time identifying and engaging these slow-moving, relatively stealthy drones.

With time, as the Russians adapted and withdrew their forces to the east, they consolidated their air-defense systems into a denser, integrated air-defense network that shortened the lifespan of Ukrainian TB2s in the battlespace. This, along with the lack of MAM-series missiles, degraded the role of the TB2 in subsequent combat but, as events unfold in Ukraine, the use of small unmanned aerial systems and loitering munitions, often called kamikaze drones, is increasing. New and inexpensive ways to manufacture these drones are also growing. Russia purchased hundreds of Shahed-136 drones from Iran in the summer of 2022 and used them to bombard Ukrainian cities. Shortly after the Russians invaded, the US sent at least 700 Switchblade loitering munitions to Ukraine[7] and, in March 2023, an Australian company, SYPAQ, provided Ukraine with small UASs, made primarily from cardboard, that use an Android cell phone as a controller. Named the Precision Payload Delivery System, this waxed-cardboard drone costs from US$680 to US$3,400 each, depending on the instrument package, and can fly autonomously, "with no operator control needed. And while it will use GPS (global-positioning system) guidance where available, if GPS is jammed the control software can work out its position from speed and heading. This means the drone can carry out missions even under conditions of complete radio jamming."[8] Cheap, disposable cardboard drones may not seem like weapons of war, but they are already combat tested by the Ukrainian Army and are conducting ISR, dropping supplies, flying as decoys to draw enemy fire, and as lethal, one-way, munitions.

Defeating Top Attack

To defeat top-attack munitions there are four general categories of technologies: laser, microwave, electronic jamming, and kinetic attack. Laser weapons use directed energy to focus a beam of light to heat up and burn through an incoming projectile or UAS. High-power microwave weapons use the power of directed microwaves to overload circuitry and fry electronic components to knock drones out of the sky. Electronic jamming interferes with the drone's control, guidance, and targeting systems. Kinetic counter-unmanned aerial systems (CUAS) employ a variety of methods to damage or destroy the drone to knock it out of the sky.

Lasers

High-energy lasers are almost ready for prime time. Almost. An example is Lockheed Martin's Advanced Test High Energy Asset, or ATHENA. The system employs a 30-kilowatt laser weapon that combines the power of three 10-kilowatt fiber lasers into one 30-kilowatt beam. This system successfully knocked out multiple rotary small UASs in a demonstration at Fort Sill, Oklahoma, in 2019. ATHENA is a transportable system, but not yet a mobile system. In 2022, Israel announced it will produce Iron Beam—a high-power solid-state laser system designed to intercept rockets, mortars, and UAVs—to supplement its Iron Dome system. Lasers require a reliable and powerful energy source, making most of these systems suitable only for the defense of fixed areas and installations. To fill the tactical capability gap for ground-maneuver forces, the US Army wants an armored, mobile, laser CUAS and contracted Raytheon Intelligence & Space in McKinney, Texas, to build and deliver three combat-capable 50-kilowatt-class high-energy laser weapon systems mounted on eight-wheel Stryker armored fighting vehicles. During the tests in 2022, the system successfully defeated several incoming mortar rounds. Three Raytheon systems, at a cost of US$123 million, is a start, but falls short of meeting the full, tactical requirement. As power-source technologies improve, the use of powerful mobile lasers to defeat UASs and incoming projectiles at the speed of light will become a reality, but fielding a viable CUAS today, rather than in 3–10 years, is a critical requirement.[9]

Microwave

Advances in microwave technology offer another option for effective CUAS weapons. An example is the Tactical High Power Operational Responder, or THOR, a high-energy microwave, laser-directed-energy weapon. THOR, developed by the US Air Force Research Laboratory, uses a focused beam of energy to counter UAS swarms. It is relocatable, but not a mobile air-defense system. As with lasers, the requirement for a reliable and high-energy power source relegates most microwave or electronic-beam weapons to protecting bases and fixed locations.

Electronic-Warfare Jamming

Electronic means to disrupt UAS operations involves the transmission of Radio Frequency (RF) signals to jam, interfere with, or take over the UAS control signal. Black Sage, an Idaho based CUAS Defense company, has developed the Goshawk Long-Range Jammer (named after a bird of prey). Goshawk is a directional non-kinetic effector that disrupts global-navigation satellite systems (GNSS) signals at a range exceeding 35 kilometers (22 miles). While transportable, it is not yet deployed in an armored and mobile configuration that could keep up with advancing tactical formations. Electronic-warfare CUASs are very promising and must be part of a layered air-defense network. Newer UASs, however, are being designed to operate without the need for GNSS and are fitted with anti-jam antennas which will make them more difficult to disrupt.

Kinetic Means

Kinetic CUAS technologies offer an immediate, reliable, and cost-effective means to counter drones in the close battle area. Kinetic systems such as DroneBullet, developed by Canadian company ArialX, employ drones to hunt and destroy drones. DroneBullet is a beyond visual line-of-sight, kamikaze counter-drone solution that defeats enemy drones by crashing into them. The system is portable, fire-and-forget, and fully autonomous. It uses on-board artificial intelligence and advanced machine-vision processing to destroy its target. Another promising kinetic system, the Raytheon Coyote Block 2+, is tube-launched, weighs 5.9 kilograms (13 pounds) and can carry a variety of interchangeable payloads, including a proximity warhead to destroy enemy drones. It has a maximum airspeed of 81 mph and a cruising speed of 63 mph. It can fly at altitudes of 30,000 feet and operate for up to an hour. The US Army recently selected the Coyote Block 3 loitering munition with a non-kinetic effector (using a microwave jammer) for its near-term CUAS solution. In a press release in 2021, Raytheon claimed the Block 3 Coyotes "engaged and defeated a swarm of 10 drones that differed in size, complexity, maneuverability and range" using a "non-kinetic effector."[10]

In January 2022, the director of US Joint Counter-Unmanned Aircraft Systems Office announced the US Army would prioritize the development of non-kinetic technologies to combat the growing threat posed by unmanned aircraft. Tests in April 2022 focused on high-powered microwave technology, directed-energy technology, and electronic warfare to counter drones. The US Army is also testing kinetic CUAS weapons. Jamming, microwave, and laser CUAS systems have advanced in recent years, and are a vital part of a fully integrated CUAS effort, but kinetic strike is also needed. As important as the proper mix of systems is, it is vital to speed up the ability to detect and defeat these top-attack weapons. To do this, a fully autonomous counter-drone system is an essential next step for CUAS.

Top Attack is a Decisive Method of Engagement

UCAVs and loitering munitions provide an inexpensive substitute for conventional air power. Any state (or non-state actor) with the resources to purchase top-attack systems on the global market has the potential to achieve drone air superiority. Winning against drones will require both masking and fielding an effective network of new CUAS weapons. The lessons of recent wars send a clear message: in the Second Nagorno-Karabakh War, and the Russia–Ukraine War, the lack of an effective CUAS capability to protect ground forces led to heavy casualties and impeded ground maneuver. On March 16, 2022, a Jamestown Foundation analysis of the effectiveness of Ukrainian TB2 UCAVs against Russian forces reported: "Of the 15 SAMs eliminated by kinetic hits, 9 platforms were targeted by Bayraktar TB2s.

Unmanned aerial systems, often called drones, are a significant threat in today's battlespace. In this photo, a counter-unmanned aerial system called the "Howler" uses a Coyote Block 2 counter-drone loitering munition and radar that identifies and defeats enemy drones. Coyote can fly individually or networked into a swarm to execute surveillance, electronic-warfare, and strike missions. (US Army photo)

All in all, the Turkish drones secured about thirty percent of the total SAM kills, and 60 percent of the direct, kinetic salvos."[11] Fielding CUAS systems and embracing masking are essential means to shield friendly forces from enemy top-attack systems. Few armies have fielded effective CUAS systems, and these are not deployed in the numbers needed to provide an adequate defense. As for the US Army, a 2021 article in *Military Review* stated this deficiency clearly: "Current Army capabilities and doctrine, especially that found in Army Techniques Publication (ATP) 3-01.81, Counter-Unmanned Aircraft System Techniques, are insufficient to meet the demands of the present and future battlefields."[12] There are many counter systems in testing, and many are promised on the horizon, but few deployed. While the US Army has identified this shortcoming, and is working the problem, there is an urgent, near-term need to provide more CUAS units and assets to defeat the drone threat.

CHAPTER FIVE

Artificial Intelligence and the Accelerating Tempo of War

> Strategy is the science of making use of space and time. I am more jealous of the latter than the former. We can always recover lost ground, but never lost time.[1]
> AUGUST GRAF VON GNEISENAU

"Jukie" is seated at his desk and looking at his computer screen. This is not his true name but his "soldier's name" and he takes pride in it. It was the nickname of his grandfather, who fought in the 1956 and 1967 wars as a paratrooper for the Israel Defense Forces (IDF). This pride runs deep in his family. In the 1973 Yom Kippur War, his uncle had fought as a tank officer. Now, Jukie, the youngest in a family of four, is serving. He has a lot to live up to, so to remind himself of his heritage, he took on his grandfather's first name as his *nom de guerre* when he enlisted in the IDF's most secret unit.

Ironically, Jukie is one of the oldest men on his team. Since he just turned 22, he feels this distinction is quite an honor. He entered the IDF when he was 17, fresh from graduating from Magshimim, a three-year after-school program for 16–18-year-old students with exceptional computer-coding and hacking skills. As one of his projects, Jukie built a smartphone from scratch. Extremely bright, the best in his class at computer algorithms, he entered college early and mastered all his computer-programming courses. This excellence brought him to the attention of IDF recruiters, but not just any recruiters. These recruiters were seeking people with special talents for an extraordinary combat unit. They offered Jukie a chance to enlist in a select team. He was eager to accept, passed the secretive entrance exams with the highest marks, and joined the IDF's elite Unit 8200.

Unit 8200 is a unique technology intelligence organization, one matched by only a few countries in the world. The unit selects young men and women for their abilities to learn and adapt to changing situations. The IDF recruiters prize those with computer-coding and hacking skills. Most of the recruits are self-taught savants. Because of the selection process of the best and brightest, and a leadership that understands how to nurture and impel exceptional young minds, Unit 8200 is among the top technical-intelligence agencies. Members who "retire" from the

Unit in their mid-20s have started some of the most successful multi-billion-dollar high-tech companies in the world. This list includes companies in the fields of cyber-security, cyber-intelligence, web-mining, software services, bioengineering, and artificial intelligence (AI). The IDF soldiers in Unit 8200 are intellectual, analytical, imaginative, and motivated.

Jukie's workstation is surrounded by a dozen other operatives sitting in front of similar computers. His office is in a secure, protected, and secret location near Tel Aviv. He can feel the energetic atmosphere in the bunker. He knows everyone is ready for action, and he feels pride in being a part of this special section at this crucial time. The young men and women near him focus on their computer screens. He closes his eyes for a moment and visualizes the next step. He is not anxious about the unfolding events. Anxious is a feeling few experience during operations in Unit 8200. Anticipatory is more accurate.

His section has practiced many simulated battles in the past year, and were involved in a dozen real cyber actions, but this time it is war. It is the responsibility of each monitor operator to oversee the AI programs that combine thousands of pieces of information about enemy activity. He remembers something one of his instructors once told him. A year ago, Jukie volunteered for "Krav Maga" hand-to-hand training. Never a very physical young man, he liked how the martial arts training gave him confidence. His instructor said, "When it comes to choke holds, there are no tough guys. Once you lock them in a choke hold, everyone goes to sleep." This is what Unit 8200 is about to do to the enemy.

He looks up for a moment and glances at one of the big displays positioned on the wall and instantly understands what is playing out in the unfolding battle. On a large screen, real-time images from one of a dozen unmanned aerial vehicles (UAVs) depict events occurring in the battlespace. On another screen, a string of Hamas missiles race up into the sky, their white contrails of smoke designating the deadly path of each rocket. Upon reaching the apogee of their ballistic trajectory, they hurtle down toward targets in Israel. Jukie watches in fascination as Israeli Iron Dome missiles launch to intercept the incoming Hamas rockets. Remarkably, Iron Dome hits every Hamas projectile in this string. Enemy rockets explode in the air in an amazing display of smoke and fire. Next time, some missiles may get through and will fall on Israeli homes, shops, and businesses. Israelis will die every time Iron Dome fails to hit its targets. It is a tribute to the competence of Israel's air defense system that so few of the Hamas missiles get past the Iron Dome, but no protection is perfect.

Jukie's section leader, an officer only a few years older, looks at the screen and orders, "Send the message."

Jukie nods and inputs the codeword to transmit a text message in Arabic to the enemy's network. A minute passes and then several icons on the screen catch his eye. Threat indicators are moving. Multiple simulations have revealed a pattern to

Hamas's reactions to an Israeli counterattack. Using the IDF's extensive database of multisource intelligence information, each key Hamas leader is tagged as a High-Value Target (HVT) or High-Pay-Off Target (HPT).[2] The AI automatically searches for HVT communications signals using patterns learned from earlier simulations, forecasts the targets' movements, and communicates with multidomain sensors to locate and track targets in real-time. As the HVTs are monitored, the AI connects multiple intelligence data points, including human, signals, geographic, satellite, and real-time video intelligence. Flying over Gaza, a swarm of drones and airborne sensors relay this data to the network which the AI automatically categorizes and sorts.

The section leader puts his hand on Jukie's shoulder. "Your analysis?"

"They are moving as anticipated," Jukie replies. "This is similar to our last battle sim."

"Excellent," the officer says, and then walks on, looking at the screens of other operatives and providing advice and encouragement.

Enemy cell-phone traffic increases. Minutes pass but seem like hours to Jukie. It takes time for the icons on his screen to move through tunnels deep underground the city of Gaza. The Hamas tunnel network is very sophisticated. Hamas has spent nearly US$1.25 billion since 2007 to build it. Despite the fact a tremendous amount of this money was stolen by corrupt officials, Hamas had enough funding to build one of the most extensive and elaborate tunnel networks in the world. They dubbed it the "Metro."

In 2016, Israel began the construction of a great wall to separate itself from Gaza. This 40-mile-long (65 kilometer) wall is nearly 20 feet tall (6 meters), six feet (2 meters) underground, and enhanced with sensors and the latest surveillance systems. This "smart fence"[3] impedes infiltration attacks and has saved countless Israeli lives. Hamas cannot break through the wall, tear it down, or climb over it. The only way to get at Israel is to dig deep under it. The Metro is Hamas's effort to tunnel under the wall, and it is an amazing feat of engineering. Composed of a spiderweb of underground approaches over 60 miles long, these infiltration tunnels lead into Israel. When they want to initiate a raid, Hamas fighters will use the Metro to come out behind Israel's defenses.

As IDF sensors detect more HVTs, new details on the computer's representation of the Metro map are displayed on Jukie's screen. The AI projects the length and depth of each tunnel using the accumulated data and reports a "confirmation rate" of 90 percent. With each passing minute, additional data is analyzed and the confirmation rate increases. Jukie observes the HVTs as they converge inside the Metro. The "fake" cell-phone message has worked.

At the same time, an extensive effort to minimize civilian casualties is underway. Gaza is a densely populated city and the IDF hopes to avoid harming civilians. Some of the advance-warning methods the IDF uses include phone and text messages to civilians and "roof knocking" with small, non-explosive projectiles.

High-flying Hermes UAVs, used for video surveillance, track non-combatants and relay real-time video images to the IDF command center.

Jukie sees the pattern as the red icons on his screen merge into an underground bunker. The enemy HVTs are gathering in "safe places" in the Metro. With this information, Unit 8200 knows the location of their hiding points. An indicator on his screen flashes, alerting him to the next action. He zeroes in on a section of the Metro as more HVTs congregate. Alerted to the situation, he turns his head to the right and announces, "Our AI is sending the message."

"A few minutes earlier than you thought, hey, Jukie?" The officer chided.

Jukie nods. He scans the reports as electronic sensors relay real-time videos.

"What is the status of our aircraft?" the officer asks.

"They are in range of the strike zone and waiting for the signal to launch effects," another operator of Jukie's section announces.

"Now we will see if what we have imagined will work," the officer replies.

Seconds pass while Jukie studies the screen. Then, the indicator icon he is waiting for flashes on his monitor, but he does not need to do anything. Multidomain strike

Top-attack systems, such as Israel Aerospace Industries' Harop pictured above, played a vital role in Azerbaijan's victory during the Second Nagorno-Karabakh War and the Israel-Hamas War (2021). The Harop is an unmanned loitering munition, with a 1,000-km (620-mile) range and a 23-kg (51-lb) warhead, that can operate autonomously during much of its operation. It can fly to a designated area, loiter for six hours, autonomously identify, categorize, and track targets, and strike autonomously. There is a mode where a human operator can abort the attack. It is highly effective in the suppression of enemy air-defense missions. (Israeli Aerospace Industries)

systems are ordered to launch. The synchronization of multidomain fires, jamming, and cyberattacks happens almost instantly. His only job is to abort the attack. The AI is running the show, he is there to cancel actions if any AI actions are deemed inappropriate. Otherwise, his purpose is to monitor and let the AI orchestrate the fight.

"The AI has issued orders to the aircraft," Jukie announces.

Jukie looks up at the large screen on the wall which depicts a multidomain common operational picture of the IDF's forces, their actions, and known enemy targets. The scene on the display shifts to a section of Gaza. The screen also depicts a real-time video, picture-in-picture stream from the high-definition thermal camera of a high-flying Hermes UAV. Unseen Israeli Air Force fighter-bombers release scores of specially designed Joint Direct Attack Munitions bunker-busting bombs and 285-pound GBU-39 Small Diameter Bombs on a dozen designated targets. Jukie watches as small explosions erupt in a line pattern on the surface of Gaza. These strikes were specifically targeted to avoid civilians. Seconds later, huge explosions burst deep underground and thick, black clouds rise from the Metro line.[4]

The operators of Unit 8200 cheer, but only for a few seconds. Focused on the job at hand, they immediately return to their video terminals to watch for indicators and determine the bomb damage assessment for each target. More aircraft fly into strike range and launch their munitions. Most of the enemy indicators turn red, flicker, and one by one, turn gray. The IDF has just fought the first "AI war."[5]

There is no transcript of Unit 8200's dialogue during Operation *Guardian of the Walls*, but the **outcome** was as depicted in this fictionalized narrative. In May 2021, the Israel–Hamas War involved new tactical methods and innovative technology on both sides. A complete military analysis of the war remains to be written, as most of the operations on both sides remain secret. A review of the available, open-source intelligence, however, offers interesting insights into the fighting and the lessons of this short conflict. As 400 rockets a day rained down on Israel, the IDF tricked Hamas leaders into bolting into the Metro on May 14, 2021, and then blasted the tunnel network with precision bunker-busting bombs. The IDF used the power of sensors, shooters, and AI to locate, track, and destroy their enemies. Fake messages tricked Hamas fighters and they reacted as the IDF's AI had predicted. Trusting in the safety of their tunnels, they moved into assembly points deep underground in the Metro. The IDF knew where they were and tracked them in real time, processing multidomain sensor and targeting data in seconds rather than minutes.[6] Once the Hamas fighters collected in clusters, the Israelis launched a devastating attack. "The IDF ground and air forces reportedly conducted a total 1,500 strikes in the 11-day war, injuring 1,900 Palestinians and killing at least 254 … The IDF claimed it killed at least 225 Hamas and Palestinian Islamic Jihad (PIJ) fighters and 25 of their leaders."[7] Operation *Guardian of the Walls* was a successful, asymmetric battle against an irregular enemy force conducted in a dense urban battlespace with

Israeli sensors, strikers, and decision makers enabled by AI. Unable to operate at the tempo of the IDF, on May 21, Hamas asked for a cease-fire.

The most important factor for Israel's successful execution of Operation *Guardian of the Walls* was the IDF's ability to create a "seamless exchange of technology, comprehensive data management, extensive defense digitalization, and a new Concept of Operations (CONOPS) called 'Intelligence-as-a-Service'."[8] AI-powered applications streamlined and analyzed a vast amount of data from a multiplicity of sources. With this exchange of data, the IDF destroyed their opponent's rocket factories, command facilities, key tunnel networks, and weapons and ammunition dumps. A vital lesson from this conflict is that digitally driven intelligence led the successful combat effort.

Developing this capability started on February 13, 2020, when the IDF Chief of Staff, Lt. Gen. Aviv Kochavi, unleashed a five-year plan called Tnufa (Momentum). Tnufa's aim was to reorganize the IDF into a smaller, smarter, multidimensional force, with AI playing a central part in military operations. This reorganization was designed to shorten engagements, accelerate the kill chain, employ big-data analysis, lessen civilian casualties on both sides, and win quickly. Big-data analysis was crucial to Israel's intelligence operations, and they used many methods, tools, and applications to collect, process, and derive intelligence data from a multiplicity of high-volume, high-velocity information sets. In short, big-data analysis reviews structured and unstructured information to generate insights, uncover hidden patterns, optimize operations, and predict future outcomes for decision makers. Generating such a high tempo makes the enemy look like a boxer standing still in the ring while the AI-enabled opponent strikes a dozen times at the most critical areas. AI, therefore, is the ultimate disrupter.

***

The tempo of war is hyper-accelerating as technology advances. Tempo is the "relative speed and rhythm of military operations over time with respect to the enemy."[9] Tempo "implies the ability to understand, decide, act, assess, and adapt … A rapid tempo can overwhelm an enemy force's ability to counter friendly actions, and it can enable friendly forces to exploit a short window of opportunity."[10] In ancient times, the tempo of operations moved at the speed of human and animal muscle power. Advances in technology increased tempo over the centuries. During World War II, the speed of the internal-combustion engine and the radio framed the tempo of operations. The pace of decision making also affected tempo as commanders could issue orders in person, by messenger, then by telegraph, and radio. Today, tempo occurs at electronic speeds as commanders issue orders using digital systems and, as evidenced in Operation *Guardian of the Walls*, when the Israelis decided and acted in real time, enabled by AI. To meet the demands of tempo, military leaders must have the mental agility to plan, prepare, and execute operations to set the conditions for success and adapt rapidly when changes are required. As war grows more complex, commanders need the help of AI to understand the battlespace and decide in real-time.

AI is the means by which humans program machines to accomplish tasks. *Encyclopedia Britannica* defines AI as "the ability of a computer or a robot controlled by a computer to do tasks that are usually done by humans because they require human intelligence and discernment. Although there are no AIs that can perform the wide variety of tasks an ordinary human can, some AIs can match humans in specific tasks."[11] There are three general categories of AI. Artificial Narrow Intelligence (ANI), or weak AI, is what we use today in our smart and intelligent systems. ANI is merely the application of computer code to sort, prioritize, and label information. We use ANI in modern military systems for a myriad of tasks including intelligence, surveillance, and reconnaissance, targeting, early warning, logistics, command and control, and to operate robotic systems. ANI can take care of the dull and dangerous work of war to change data into useful warfighting information and remove humans from dangerous tasks. For the rest of this book, ANI will simply be called AI.

Artificial General Intelligence (AGI), strong AI, is another matter. We do not have AGI today, but we are getting close. AGI is the hypothetical ability of a computer to operate at the cognitive level of a human mind. There is much debate in the scientific computer world on the possibility and wisdom of creating AGI. When you think AGI, you are edging toward "Terminator"-like abilities, from the classic 1984 movie of the same name. No nation has yet produced AGI, but many are working to develop it. Artificial Super Intelligence (ASI), also a strong form of AI, is the next level and a transcendent capability. Happily, at the time of this writing in 2023, AGI and ASI only exist in the realm of science fiction. The fear, however, is that what humans can visualize, even in science fiction, can become reality. Consider the communicator of the 1965 television series *Star Trek* with the development of today's smartphone.

AI has three subsets: machine learning (ML), neural networks, and deep learning (DL). ML is the science of applying AI to provide machines with data and then allowing them to learn for themselves. A neural network is a subset of ML and is defined as a "series of algorithms that endeavors to recognize underlying relationships in a set of data through a process that mimics the way the human brain operates."[12] You must have a neural network to generate DL. Microsoft defines DL as:

> ... a type of machine learning that uses artificial neural networks to enable digital systems to learn and make decisions based on unstructured, unlabeled data. In general, machine learning trains AI systems to learn from acquired experiences with data, recognize patterns, make recommendations, and adapt. With DL in particular, instead of just responding to sets of rules, digital systems build knowledge from examples and then use that knowledge to react, behave, and perform like humans.

As anyone can see, AI is changing the way we live, work, and fight. AI can intelligently sort through large amounts of data to discover optimal targeting parameters much

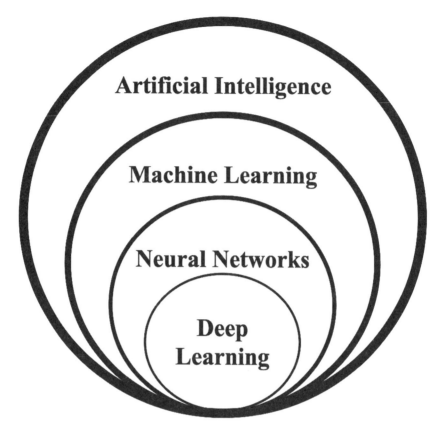

Machine learning (ML), neural networks, and deep learning (DL) are the three main subgroups of artificial intelligence (AI). (Author's image)

faster than any human. As a result, advances in AI, ML, neural networks, and DL are creating new military capabilities that are disrupting the methods of warfare.

An everyday example of AI is the generative AI app ChatGPT (Chat Generative Pre-trained Transformer). Available for free on the internet, ChatGPT is a complex machine-learning model, developed by OpenAI, a research firm co-founded by Elon Musk. Generative AI is "a broad label that's used to describe any type of artificial intelligence (AI) that can create new text, images, video, audio, code or synthetic data."[13] I asked ChatGPT the following question: "How is China using AI for its military?" Below is the detailed answer generated by ChatGPT:

> China has been investing heavily in AI technology for its military in recent years, with the aim of creating a world-class military that can compete with other major powers. Here are some ways in which China is using AI for its military:
> 1. Autonomous Weapons: China is investing in developing autonomous weapons, such as unmanned aerial vehicles (UAVs) and ground-based robots, that can operate without human intervention. These weapons are equipped with AI algorithms that allow them to analyze data and make decisions in real-time.

2. Command and Control: China is using AI to improve its command and control systems, which allow commanders to make better decisions on the battlefield. These systems use AI algorithms to analyze large amounts of data and provide real-time recommendations to commanders.
3. Cybersecurity: China is using AI to enhance its cybersecurity capabilities. AI algorithms are used to monitor network traffic.[14]

As you can see, this is an impressive answer for a weak AI language program. ChatGPT created this answer in less than a minute. Imagine how improved versions of applications like ChatGPT will aid military decision makers in the years to come.

The Race Goes to the Swift

Time is essential in war. Winning time wins battle. Observing and acting too late is a sure recipe for defeat. Admiral Lord Nelson, the victor of the epic naval battle of Trafalgar on October 21, 1805, that established British naval supremacy for more than a century, once said, "Time is everything; five minutes make the difference between victory and defeat."[15] Today we calculate this margin in seconds and micro-seconds. A case in point is what Operation *Guardian of the Walls* achieved in 2021. Speeding up the ability to sort, prioritize, recognize patterns, and act on this accumulated information in seconds, rather than hours, makes AI an essential tool of modern warfare. As we connect more military systems in AI-enabled networks where weapons systems transmit and share information, AI will sort through thousands of data points, correlating their significance, recognizing the patterns, and providing battle commanders with actionable courses of action. The military that uses AI to synchronize multidomain kinetic and non-kinetic effects at machine speeds will gain a significant advantage over those who do not. This is the essence of war in the 21st century.

Data is now a weapon. Turning data into information and leveraging information at scale in near real-time requires robust AI. Connecting AI with networked weapons will accelerate the orient, observe, decide, and act loop, and generate speed in decision making and execution. The goal is for AI to think and act faster than any adversary; Christian Brose, the best-selling author of the book *Kill Chain: Defending America in the Future of High-Tech Warfare*, emphasized the importance of AI in his testimony to the Congressional Armed Services Committee on February 9, 2023:

> In the recent Nagorno-Karabakh conflict, in the continued fighting in the Middle East, and in the ongoing war in Ukraine, we are seeing how low-cost robotic vehicles, AI-enabled loitering munitions, digital targeting systems, cyber weapons, persistent communications and surveillance satellites, and other advanced capabilities—especially when paired with large volumes of more traditional weapons—are transforming the modern battlefield.[16]

AI can synchronize the effects of loitering munitions, long-range precision fires, sensors, and a host of robotic vehicles. AI is the greatest disrupter of our time and will have a dramatic effect on the conduct of war.

The US military has not faced a peer in combat since World War II. Although it faced a peer enemy in Soviet Russia during the 54 years of the Cold War, the US was not put to the ultimate test. During World War II, the greatest technological race was to develop the atomic bomb. The challenge we face with AI is similar today. In 1942, the US began the *Manhattan Project*, the codename for the effort to develop an atomic device. Then, there were many skeptics that said it could not be done. Germany was working on a similar project, but did not invest the focus, talent, energy, and capital to succeed. The US invested about twenty billion dollars (in 1996 dollars) and won the race.[17] This decision helped to make the US a superpower and set the course for the next 80 years. America's standing in the world today derives from the World War II generation's efforts to do what needed to be done, no matter the risks or cost.

Today, China is determined to become a technological superpower and believes AI is the strategic technology leading the future. Central to this effort is the New Generation Artificial Intelligence Development Plan.[18] This 28-page document is an effort that is like the *Manhattan Project* and is driven by China's Ministry of Science and Technology. The goal of the plan is to make China the dominant AI power before 2030. This goal should concern every US and Western leader.

China has the world's largest military, the second largest economy, and its technology industry is rapidly advancing. AI development is at the center of

Semi-autonomous	Human-supervised	Fully Autonomous
"Human in the loop"	**"Human on the loop"**	**"Human out of the loop"**
Weapon system that, once activated, is intended to **only engage individual targets or specific target groups** that have been selected by a human operator. Includes "fire and forget" munitions	An autonomous weapon system that is designed to **provide human operators with the ability to intervene and terminate engagements,** including in the event of a weapon system failure, before unacceptable levels of damage occur.	A weapon system that, once activated, **can select and engage targets without further intervention by a human operator.**

(Note: Definitions per DoDD 3000.09, *Autonomy in Weapon Systems*)

According to DoD Directive 3000.09, autonomy in weapon systems is defined as a "weapon system that, once activated, can select and engage targets without further intervention by a human operator." This concept of autonomy is also known as "human out of the loop" or "full autonomy." (US Department of Defense photo)

Developments in Artificial Intelligence will change both fire and maneuver in the future battlespace. In this photo, US Army troopers assigned to the 3rd Armored Brigade Combat Team, 1st Cavalry Division, fire the M1A2 SEPV3 Abrams main battle tank as part of gunnery qualification on September 22, 2022, at the Mielno Tank Range, Drawsko Pomorskie Training Area, Poland. (US Army photo by Staff Sgt Charles Porter)

this effort. China leads the world in AI patents and publications, AI-empowered speech and image recognition, 5G, and drone manufacturing. In 2023, only 60 percent of China's population used the internet, compared to 89% in the US, but "its sheer scale means there are three times more internet users than in the United States—over 800 million. Nearly all of China's internet users access the web through mobile devices from the country's biggest smartphone companies, Huawei, Oppo, and Vivo."[19] Most critically, according to the Australian Strategic Policy Institute (ASPI), an independent technology think tank, China leads the world in 37 out of 44 critical technologies. The US leads in only seven technologies on ASPI's list—vaccines, semiconductors, high-performance computing, advanced integrated circuit design, natural language processing, quantum computing, and space launch systems—and comes in second in most of the other critical technology development areas. "A key area in which China excels is defense and space-related technologies. China's strides in nuclear-capable hypersonic missiles reportedly took US intelligence by surprise in August 2021 … These findings should be a wake-up call for democratic nations, who must rapidly pursue a strategic critical technology step-up."[20] The Chinese have advanced their understanding and development of AI in a very short amount of time. The US military recognizes AI is the key to the future of warfare, but so does Russia and China. Soon, they may put us to the test. Clearly, the AI race is to the swift and the US and the West are falling behind.

The Sierra Nevada Corporation, Voly-T is a launch system-independent, unmanned, reconnaissance, surveillance, and target acquisition system. Enabled by Artificial Intelligence (AI) drones are able to self-organize, recognize the environment, identify threats, locate targets, and attack without direct human control. (US Army photo)

CHAPTER SIX

The Transition to Fully Autonomous Weapons

> In future battlegrounds, there will be no people fighting ... Mechanized equipment is just like the hand of the human body. In future intelligent wars, AI systems will be just like the brain of the human body ... Intelligence supremacy will be the core of future warfare and AI may completely change the current command structure, which is dominated by humans, to one that is dominated by an AI cluster.[1]
> ZENG YI, A SENIOR EXECUTIVE AT CHINA'S NORINCO DEFENSE COMPANY, THE NINTH LARGEST DEFENSE COMPANY IN THE WORLD

Imagine a peer conflict in 2030. Warfare is hyper-accelerated. Sensors blanket the battlespace. If you are seen, or emit an electronic signature, the enemy locates and hits you. Networked, smart munitions are becoming increasingly intelligent. These weapons, along with non-kinetic effects such as cyberwarfare and electronic warfare, when linked to a multidomain sensor network, and enabled by artificial intelligence (AI), and its subsets, machine learning (ML) and deep learning (DL), dominate the battlespace. ML allows the AI to automatically improve through experience. DL imitates the workings of the human brain in processing data and creating patterns to select decisions. AI and its subsets, referred to simply as AI, will operate as a "neural network" forming a "brilliant"-fires system that will deliver extraordinarily accurate, extremely long-range, and exceptionally lethal, fires in real-time.

Technology is developing at a pace where many systems are beyond human cognition to control. To adapt, militaries are moving to weapons with pre-programmed decision-making capabilities and more fully autonomous weapon systems. The combination of AI with long-range precision fires (LRPF) will be revolutionary. LRPFs, which include many types of delivery systems (artillery, drones, balloons, and missiles launched from ships and aircraft) linked with multidomain sensor networks, and synchronized and optimized by an AI neural network, will generate a super-fast kill web to accelerate the lethality of war. This neural network will enable a kill web that will act as a weaponized "Internet of Battlespace Things." Such a brilliant-fires system could make maneuver in the physical domains of war (land, sea, air, and space) very difficult, if not impossible. With such a capability on both sides of a conflict, the side that strikes first will gain a significant advantage as the opponent's systems are rapidly eliminated in the first wave of this brilliant-fires assault.

Unable to maneuver in such a deadly battlespace, the contest could rapidly become deadlocked in an ever-decreasing exchange of missiles until, eventually, one side exhausts its fires. As seen in the fighting in Ukraine, the depth of the ammunition magazine matters. Potentially, this could force a peer adversary to employ nuclear weapons or face complete annihilation by the side that has more precision weapons. This could all happen very quickly in the opening hours of a conflict.

The power of cognitive technologies is being driven by advances in AI. As mentioned previously, ChatGPT stunned the world in December 2022, with its intelligent answers and detailed, human-like responses. Dall-E, also created by OpenAI, is an AI program that draws realistic digital images from text or voice description.[2] Militaries have always experienced evolutionary pressures from war, but the current technological pace of AI development in the commercial world is accelerating at a pace greater than anything yet experienced. Moore's Law predicts computer processing speeds will double every 18 months to two years. A Stanford University study revealed that AI computational power is accelerating faster than Moore's Law.[3] If this rate of acceleration holds, AI will be able to do extraordinary things by 2030.

One way to visualize advances in technology in the years ahead is the Gartner Hype Cycle, a graphical model that depicts the maturity and adoption of new and emerging technologies. It is a running projection of technology trends that tracks the introduction to maturity of key technologies. Aimed at a technology's relevance in solving problems, the Hype Cycle can assist leaders regarding technology adoption. For instance, the 2022 Hype Cycle for artificial intelligence predicts decision intelligence (decision making to design, model, align, execute, monitor, and tune decision models and processes) will attain a plateau of productivity in 2–5 years, smart robotics in 5–10 years, and autonomous vehicles in more than 10 years. Even if those prognostications are off by a few years, the trend is for rapid technological growth. Futurist Ray Kurzweil, who has been a Director of Engineering at Google since 2012, explained, "Humans are wired to expect linear change from their world. They have a hard time grasping the accelerating exponential change that is the nature of information technology."[4] The Hype Cycle is one tool to help leaders grasp that exponential change.

The greatest changes in military technology in the next decade are predicted by the Hype Cycle to involve data fusion, visualization, and robotics. The US Army is developing data fusion and visualization tools using the systems names JADC2 and TITAN (Joint All-Domain Command and Control and Tactical Intelligence Targeting Access Node, see Chapter 7). Work on robots, that fly, roll, and walk is also accelerating. In 2019, the US Army Science Board's study on Multidomain Operations, recommended the US Army "will need to develop and field optimized

THE TRANSITION TO FULLY AUTONOMOUS WEAPONS • 69

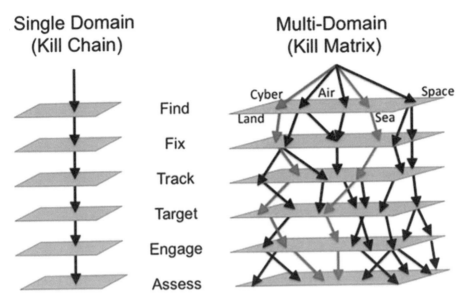

This image depicts the kill chain compared to a multidomain kill matrix. The kill-matrix concept is a start point for an artificial intelligence-enabled kill web. The web will use AI to process and operate sense-and-strike operations at machine speeds. (Army Science Board photo)

human-machine systems for the battlefield and encourage the development of the next generation of capabilities."[5] In 2030 and beyond, as the fusion of AI, LRPFs, and multidomain sensor networks matures, the requirement to enhance the survival of maneuver systems and make them more lethal at the point of decision will create a desperate demand for new capabilities. These capabilities will involve technologies to counter the ascendency of this brilliant-fires complex. The two most important capabilities to counter this emerging threat are the military concepts of masking and multidomain convergence. Masking will keep ground systems alive by making them difficult to identify and target. Multidomain convergence will make them more lethal than ever before.

Masking Against Fully Autonomous Weapons

Masking will be crucial to defeat tomorrow's fully autonomous weapon systems. Masking is more than camouflage and stealth. It employs next-generation active and passive means to reduce the electromagnetic-spectrum signature to render the system difficult to locate and hard to target. Masking includes, but is not limited to, technologies that enable spectrum management and the low probability of detection in the electromagnetic, acoustic, seismic, and thermal spectrums; intelligent, multispectral camouflage systems; decoys and spoofs; cognitive electronic-warfare systems employing ML to counter the enemy's radars; and the use of jamming, electronic countermeasures, and digital radio frequency memory to hide and operate beneath the blanket of enemy or friendly jamming.

The intention of masking is not invisibility from the unblinking eye of fully autonomous systems, although that would be ideal, but the ability to reduce the probability of detection, confuse the enemy's targeting system with false readings, and make them shoot at ghosts they cannot accurately locate. Sun Tzu, the ancient sage of war, understood masking when he stated the most refined form you can give your troops is to make them "as hard to know as shadows. They are like rolling thunder …"[6] The ability to mask ground systems, and turn them into shadows, would be a revolutionary shift in today's lethal battlespace.

Multidomain Convergence becomes All Domain Operations

Masking, as vital as it is to survive in a battlespace dominated by LRPFs and ubiquitous sensor networks, is primarily defensive. To restore maneuver to a battlefield dominated by brilliant fires requires the ability to project striking power. Without maneuver, there is no decision, only stalemate. To maneuver in this future battlespace, we must conceptualize a mobile striking power system with new capabilities that emphasize the technological-cognitive advances that will develop in the coming years. This evolutionary pressure, accelerated by exponential technological development, will also force military leaders to conceive new and better ways to execute combined arms.[7] The US Army has named this new approach "Multidomain Operations Convergence." Convergence is one of the three tenets of the Multidomain Operations (MDO) concept. It is leading the revolutionary warfighting shift in the near future that will become the new combined-arms doctrine.

The US military defines convergence as the rapid and continuous integration of capabilities in all domains. Convergence will rapidly evolve to require a fusion of our current understanding of convergence with AI to synchronize and optimize cross-domain synergy and multiple forms of attack in real-time. Since the ability to do this in real-time is beyond human capacity, AI will provide the answer. By the 2030s, MDO convergence will potentially become "All Domain Operations." ADO will be defined as the rapid and continuous integration and synchronization of all multidomain capabilities by an AI neural network that is delivered by cognitive, enhanced systems to the warfighter (a human on the loop) who is enabled by Distributed Mission Command and disciplined initiative to execute military operations.[8] Distributed Mission is the execution of Mission Command using smaller distributed command nodes to execute the functions of the command post without staff co-location. The goal of distributed mission command is to enhance continuity and survivability of the command function in a transparent and lethal battlespace. (Author's definition).

Just as people use their smartphones to navigate instead of legacy paper maps, AI and cognitive-enhanced systems will provide a soldier the ability to execute combined-arms synergy more rapidly and at a lower level of command. The primacy of AI to execute MDO was expressed in a May 10, 2019, *Naval Proceedings* magazine article titled "Operationalizing Artificial Intelligence for Multidomain Operations: a First Look": "AI/ML is a foundational requirement for Multidomain Operations (MDO) ... the future force requires the ability to converge capabilities from across multiple domains at speeds and scales beyond human cognitive abilities."[9] AI, therefore, is a fundamental requirement for ADO, as humans cannot manage a plethora of diverse capabilities without the cognitive enhancement provided by AI interfaces. AI and sensor-network fusion will drive the military to change and to respond effectively to multidomain warfare threats. Battle leaders, empowered by cognitive-enhanced systems that assist them in delivering convergence at the point of decision, will facilitate the employment of the best human warfighting capabilities to execute Mission Command like never before.

Imagine the ability to execute a multidomain kill web as a force penetrates the enemy's anti-access/area-denial zone. The advancing ground force is masked and very

Today, the US military uses a kill chain that takes minutes to execute. A kill web employs an artificial intelligence-enabled process to rapidly synchronize the effects of many networked munitions in time, space, and purpose and, therefore, speed up sensor-to-shooter timing exponentially. In this photo, US Marines fire a High-Mobility Artillery Rocket System at Camp Shorab in Helmand Province, Afghanistan, February 22, 2019. (US Marine Corps photo by Sgt Victoria Ross)

difficult to identify and target. The enemy cannot apply his brilliant fires as planned. As the masked ground force advances, it generates a strike zone to destroy targets identified to its front and flanks. This strike zone grows in front of the masked ground force as it advances, as far out as 50 kilometers (31 miles) or more, depending on the capability of the sensor network and the cognitive-enhanced interface linkage with the kill web. If sensor networks are interrupted or fail, the advancing force becomes the node that restores the sensor network, allowing other intelligent systems within range to link with it. Within that strike zone, identified enemy targets are hit and eliminated by every smart and intelligent multidomain weapon system in range, using the combined effects of capabilities from air, sea, land, cyber, and space-based platforms. By bringing the kill web down to the point of decision, ADO and masking restores maneuver, creates a means to generate a rapid decision to finish

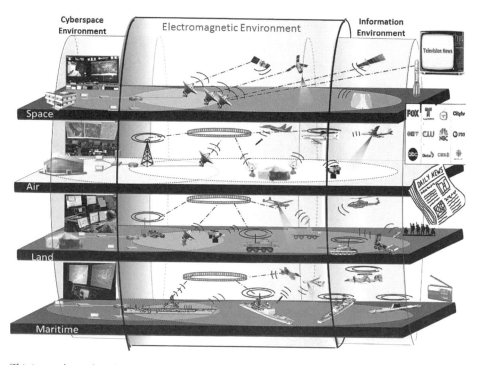

This image shows the relationship between the five domains of war (land, sea, air, space, and cyber), the electromagnetic spectrum, and information warfare. Massing the effects from one domain to another is called cross-domain maneuver. Overwhelming one enemy domain with the effects from multiple domains can generate battleshock. The cognitive and autonomous levels of the latest weapons will increase the tempo, precision, and lethality of war by linking more sensors and shooters. Gathering, analyzing, organizing, sharing, and synchronizing this volume of time-sensitive multidomain targeting data is no longer effective at human decision-making speeds. An artificial intelligence-enabled kill web is a solution to this problem. (US Department of Defense photo)

the conflict before the use of nuclear weapons, and negates the costly investments in our opponent's LRPF forces.

General Eric Shinseki, a past Chief of Staff of the US Army, famously said: "If you don't like change, you're going to like irrelevance even less."[10] In 10 years, the world will experience exponential technological change that will touch nearly every aspect of life. Disparate elements across the technological spectrum are uniting to drive expeditious changes in the development of military systems. Advances in AI are leading this transition and, by the 2030s, AI will be embedded in everything, including all state-of-the-art weapons systems. The conduct of warfare will transform dramatically as AI changes key facets and the speed of combat. The major impact of AI and multidomain sensor networks on future war will not be to create "Terminator"-like robots, but to connect and command every smart and intelligent weapon system and sensor within range of the AI neural network to form a brilliant kill web. Current trends in the development of LRPFs, extensive mesh sensor networks, and AI promise to dominate the future battlefield and render maneuver costly, difficult, or impossible. Understanding and preparing for this change takes foresight that can only be gained by comprehending the nature of war and studying the current technological trends. What we need is a merging of Sun Tzu's philosophy and the Gartner Hype Cycle. Today, military technology is experiencing this exponential change. Military analysts and decision makers should focus on these trends and looming possibilities to avoid becoming irrelevant. Masking is the shield; ADO is the sword. Fire without maneuver is indecisive. Maneuver without fire is fatal. AI will enable both fire and maneuver. Developing innovative combat-vehicle designs that include masking with AI cognitive systems to enable ADO, and integrating these into an all-domain kill web as we will see in the next chapter, will offer a revolutionary means to survive, fight, and win.

The Textron Systems - Aerosonde HQ 4.8 is a vertical takeoff and landing Unmanned Aerial System (UAS) under consideration by the US Army. Systems like this will provide Intelligence Surveillance and Reconnaissance (ISR), Electronic Warfare (EW), and other vital missions to support the US Army's kill chain and emerging kill web. (US Army photo)

CHAPTER SEVEN

The Kill Web

> What's interesting to me about the kill web concept is that our forces don't need to go looking for targets all the time. Like a spider, we set up the web around a battle space and wait for the enemy to show up. Once they enter the kill web, they can't get out.[1]
>
> RAY ALDERMAN

A 40-year-old Ukrainian lieutenant colonel walks down a path through a forest. Two armed soldiers lead the way.

The Ukrainians are on a relentless hunt, and their prey is a command post (CP). The Ukrainian commander, a long-time soldier, wants the name Chornobaivka burned into the hearts and souls of the cursed invaders. He wants the Russians to remember Chornobaivka.

His guides take him to a bunker hidden among the fir trees. He walks with them nearly the length of three soccer fields, finally arriving at the CP's entrance. Walking to the CP is the standard procedure. This helps to keep the Ukrainian CP masked from Russian sensors. Vehicles are never allowed anywhere nearby. He learned long ago that Russian unmanned aerial systems (UAS), like vultures, will follow vehicles, see where they are parked, and then target the Ukrainian CP. Headquarters are high-value targets in this war and carelessness will kill you.

He stops as the soldiers walking with him open a camouflaged entrance and lift the heavy metal covering, just enough for the commander to enter. From the surface, the entrance to the CP looks like any other patch of ground. He walks down the wooden steps, turns a sharp corner in the dimly lit passage, and enters the bunker. Inside the large subterranean room are generators, lights, maps, computers, a Starlink receiver, and six Ukrainian soldiers.

"Commander, you arrived just in time. Our drone confirms the targeting information," a young Ukrainian soldier from the Aerorozvidka unit announces. "The Orcs are very sloppy today. Bayraktar has found them."

The commander looks at the screen approvingly. He watches the real-time video feed from the Bayraktar TB2 unmanned combat aerial vehicle (UCAV) that is surveilling the Russian CP. On the monitor, he sees two Russian officers exit an

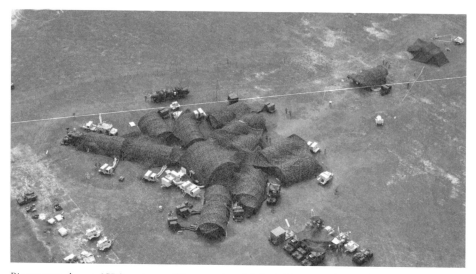

Big command posts (CPs) are impossible to mask and defend. The target that sticks out gets hammered. A CP like this is a target waiting to happen. Tents offer no protection. In the Second Nagorno-Karabakh War, the Azerbaijanis located and destroyed Armenian CPs with loitering munitions, unmanned combat aerial vehicles, and long-range precision fires. (US Army photo)

armored car and move toward a concrete-walled factory building. He knows the Russians are officers as two soldiers standing guard at the entrance salute as they approach. The officers return the salute and enter. It is clear to him the invaders have no idea they are being watched, no knowledge they are the prey. He smiles, knowing they will get a brief and final education in the next few minutes.

"Big fish have arrived," the Aerorozvidka soldier replies with a smile. He then types into a laptop computer to coordinate a missile strike with a HIMARs (M142 High Mobility Artillery Rocket System) fire-direction officer. "Fires are ready, sir,"

"Bayraktar is observing to confirm the results of the strike," a drone operator reports.

"It's time," the commander replies. "Execute when ready."

A minute passes. The commander stands quietly behind the drone operator.

"Missile to strike in five seconds," a soldier in the bunker announces.

The time passes as the Russian CP is shown in high definition on the monitor. Everyone waits and watches. Suddenly, a missile strikes the top of the factory, and the Russian CP explodes in a ball of fire and flying debris.

Everyone inside the bunker cheers. They know very few invaders will survive that blast. Ten seconds later, for good measure, a second HIMARs missile strikes the same target.

"Congratulations," the commander proclaims. "We have just smashed the headquarters of Russia's 49th Combined Arms Army. May they all rot in hell. *Slava Ukraini!*"

***

The relentless destruction of enemy command and control is intrinsic to modern war. As stated before, modern war is a shot to the head, not a bullet in the arm. On April 22, 2022, near Kherson, Ukrainian sensors positioned within range, located the forward CP of the Russian 49th Combined Arms Army (CAA), subsequently targeting it with a barrage of missiles. The story above dramatizes the action. The devastating strike killed two Russian generals, including the commander of the 49th CAA. The *Associated Press* reported "Oleksiy Arestovych, an adviser to Ukrainian President Volodymyr Zelensky, said in an online interview that 50 senior Russian officers were in the command center when it came under attack. He said, 'Their fate was unknown.'"[2] On April 30, Ukraine hit another Russian CP in Izyum, and Ukrainian sources reported the strike destroyed 30 vehicles. Ukraine reported the Russian Chief of the General Staff, General Valery Gerasimov, was flown out of Izyum with shrapnel wounds. On May 18, the Ukrainian military reported that its artillery destroyed yet another CP, this time of the Russian Black Sea Fleet. Three officers were killed and 14 were seriously injured. Ukrainian forces have targeted Russian CPs and leadership with precision weapons from the start of the war. Russia has lost over 120 CPs to these strikes. The loss of so many critical staff officers and special command-and-control vehicles would be debilitating to any army but, given the tight, centralized command-and-control philosophy of the Russian Army, these losses are disastrous. The Ukrainian kill chain proved to be very effective. "To me, and in conversations with other officers across various services," said Lt. Gen. Karsten Heckl, head of United States Marine Corps Combat Development Command, in May 2022, "clearly the ubiquity and proliferation of sensors and the ability to close kill chains accurately, precisely on any target is a major lesson to take away."[3] Robotic sense-and-strike weapons are changing the nature of combined arms, increasing the tempo and precision of war, and accelerating the kill chain.

The Kill Chain

Speed is essential to wage war successfully. Operating faster than your opponent provides a winning advantage. Currently, all military forces operate human-centric kill chains. A kill chain is a sequential process that operates at human speeds and represents the sequence of events required to sense and strike enemy targets. Traditionally human-in-the-loop- or human-on-the-loop-centered, this involved identifying a target, deploying a weapon system to engage the target, launching the munition, destroying the target, and verifying battle damage after the strike: find, fix, fire, finish, and feedback. Christian Brose, in his book *The Kill Chain*,[4] identified a three-step process: gain understanding of what is happening, decide what to do about it, and take action (kinetic or non-kinetic). The traditional kill chain can be slow, as the targeting system must pass through multiple human operators before striking power is activated. At the very best, it is measured in minutes. Experimenting with

Large command posts in tents, like this one, provide a tempting target. Imagine if a loitering munition hit this command post. The collected command and staff "brain power" of an entire armored division would be lost. We must mask our command posts and practice "distributed mission command."

new technology and procedures in September 2020 at the Yuma Proving Grounds in Arizona, the US Army was able to demonstrate a kill-chain time of 20 seconds.[5]

During the Second Nagorno-Karabakh War, the Azerbaijanis accelerated their kill chain by using sense-and-strike UCAVs, such as the Turkish-made TB2, and loitering munitions such as the Israeli-made Harop and Orbiter. With the Azerbaijani kill chain, they often measured the time from target identification to verification of destruction in seconds. The Azerbaijanis accomplished this by designating a "three-dimensional strike zone" in the battlespace, essentially a "free-fire box," to find and strike any targets according to a designated priority of targets. The target priorities executed by Azerbaijan were, in order: air defense, electronic warfare, command and control, artillery, tanks, armored vehicles, wheeled vehicles, and, finally, troops. By the middle of October 2020, the Azerbaijanis had been so successful their UCAVs and loitering munitions were hunting troops. What was equally impressive was that the high-end, Israeli-made loitering munitions communicated with each other automatically and, when a pack of loitering munitions struck their targets, no two hit the same one. This streamlined kill chain is one reason for Azerbaijan's decisive victory over Armenia.

Every kill chain involves the sequential process previously described, and each of these steps operate at different speeds. The time involved to execute the combined kill chain is the sum of its parts. In war, extraordinary speed is of great importance. Human interaction across the process slows the kill chain. As more weapons send data to each other during a strike, as the Israeli-made loitering munitions can with real-time streaming video, an Internet of Battlespace Things (IOBT) will evolve. Connecting these fast-moving systems and synchronizing their attack is beyond the means of a human-centric kill chain. In addition, the process is further accelerated if capabilities from all domains, land, sea, air, space, and cyber, can be activated as part of the kill chain. This AI-enabled, multidomain system is the emerging kill web.

The Kill Web

A kill web is an AI-enabled kill chain that connects sensors and shooters to automatically execute targeting at machine speed. The AI will synchronize the effects of many networked munitions in time, space, and purpose and will speed up sensor-to-shooter timing exponentially. In 2019, US Missile Defense Agency Director Vice Admiral Jon Hill explained the need for an AI-enabled kill web:

Army Capt. Eric Tatum, assigned to the Army Futures Command's Artificial Intelligence Integration Center, conducts field testing with the Inspired Flight 3 Drone during Project *Convergence 2022* at Fort Irwin, California, on October 27, 2022. The project's experimentation incorporates technologies and concepts from all services, and from multinational partners, including in the areas of autonomy, augmented reality, tactical communications, advanced manufacturing, unmanned aerial systems, and long-range fires. (US Army photo)

"With the kind of speeds that we're dealing with today, that kind of reaction time that we have to have today, there's no other answer other than to leverage artificial intelligence."[6] China's military experts predict lethal "intelligentized"[7] weapons will be common by 2025. The Chinese view of intelligentized warfare envisions the use of AI to enhance its military capabilities.

An AI-enabled kill web will transform warfare. Understanding this, and unwilling to lose the race for supremacy, the United States, China, Russia, Turkey, Israel, and several other nations are working on AI-enabled, multidomain kill webs to integrate and synchronize all sensors and fires. The goal of creating a system of connected sensors and weapons, communicating vital data to each other in real-time, and controlled by AI, is a daunting challenge. The US spends more on defense than Russia and China combined and has been the master of the kill chain for the past several decades, but its stove-piped, single branch (Army, Navy, Marines, Air Force, and Space Force) systems are difficult to mesh into an integrated kill web. Ray Alderman, of Vita Technologies, talked about this conundrum in a May 30, 2018, article:

> The Army has the weapons, OODA loop (Observe, Orient, Decide, and Act), and kill chain for land-based problems. The Navy has the weapons, OODA loop, and kill chain for sea-based problems. The Air Force has the weapons, OODA loop, and kill chain for air-based problems. There's also space, cyberspace, and electronic warfare (the electromagnetic spectrum) with their own weapons, OODA loops, and kill chains … If hooking all our military platforms together is a rat's nest of problems, how can we hook our allies' platforms into our network? We need to replace these two-dimensional kill chains (the static linear sequence of events) with a six-dimensional kill web (connect all six domains of warfare into a dynamic network). How do we do that? By communicating targeting data collected in one domain to all the other domains instantly, creating "shared situational awareness."[8]

Joining these different kill chains is the purpose of Joint All-Domain Command and Control (JADC2). The overall goal of JADC2 and Tactical Intelligence Targeting Access Node (TITAN) is to operationalize sensor and intelligence data, connect and synchronize every sensor and shooter by a military IOBT, and automate data at machine speeds to generate a near real-time visualization of the battlespace for warfighters. TITAN is a program to work with JADC2 to access multidomain sensors and fires and uses AI to accelerate and prioritize targeting. TITAN will synthesize multidomain sensor information and transform battlespace intelligence into targeting information for long-range precision fires. How quickly JADC2 and TITAN will do so, and how soon, remains to be seen. The bottom line is that creating an AI-enabled, multidomain kill web is an extremely challenging task that requires determined leadership, intense focus, and the combined efforts of the Department of Defense and its industry partners.

A recent example of an emerging kill web occurred during Israel's execution of Operation *Guardian of the Walls* against Hamas in May 2021. Israel declared this was the first "Artificial Intelligence War." Israel exists in a military situation where it cannot trade space for time, as it has no ground to give. Thus, it must act with alacrity in any conflict. To gain a time advantage over its opponents, and to apply lessons learned from previous wars, Israel has prioritized the development of AI and machine learning for military operations, especially targeting. The Israeli AI-enabled kill web correlated huge amounts of data, from all forms of intelligence and surveillance inputs. Hamas started the war by launching a non-stop rocket barrage against Israel on May 10, 2021. As the Iron Dome anti-missile system attempted to protect Israel from Hamas's arsenal of 190,000 rockets, Israel struck back by hitting 1,500 targets with speed and precision:

> Massive AI machinery for Big Data Analytics provided support at every level—from raw data collection and interception, data research and analysis, right up to strategic planning—with the objective of enhancing and accelerating the entire process, from decision-making about prospective targets to the actual carrying out of attacks by pilots from F-35 cockpits. These targets included Hamas rocket launch sites, command-and-control centers, weapons storage sites, and tunnel systems.[9]

On May 12, IDF strikes killed 16 key Hamas leaders. "For the first time, artificial intelligence was a key component and power multiplier in fighting the enemy," reported an Israel Defense Forces (IDF) Intelligence Corps officer on May 27. "This is a first-of-its-kind campaign for the IDF. We implemented new methods of operation and used technological developments that were a force multiplier for the entire IDF."[10] The Israeli-made kill web is a combination of AI programs, with codenames such as *Alchemist, Gospel,* and *Depth of Wisdom*.[11] These AI programs dramatically reduced Israeli casualties and ended the conflict in 11 days, with Hamas asking for a cease-fire. In an after-action report conducted by the IDF, Chief of Staff Lt. Gen. Aviv Kohavi said the IDF achieved a new level of connection between sensors, intelligence, and shooters: "During three days of Operation *Guardian of the Walls*, the Southern Command and the Gaza Division destroyed 70 multi-barreled launchers. They destroyed a multi-barreled launcher every hour."

Data is the ammunition of the next war. Multidomain sensors can collect mountains of data. Processing this data in time is beyond human cognition. Using AI to rapidly process data, which is what AI does best, can dramatically truncate the sensor-to-shooter time. The objective of the kill web is to use AI to synchronize all available weapons within striking range to engage priority enemy targets with precisely the right weapons while executing the decision making of this complex allocation at machine speeds:

> The kill chain is a line concept which is about connecting assets to deliver fire power; the kill web is about distributed operations and the ability of force packages or modular task forces to deliver force dominance in a specific area of interest. It is about building integration from the ground up so that forces can work seamlessly together through multiple networks, operating at the point of interest.[12]

The 11th Armored Cavalry Regiment and the Army Threat Systems Management Office facilitate a drone swarm of 40 drones during a training exercise at the National Training Center, Fort Irwin, California, on May 8, 2019. This exercise was the first of many held at the Center. (US Army photo by Pv2 James Newsome)

The goal is to create an AI-enabled kill web that connects all smart weapons and multidomain effects with the greatest speed, range, and decision dominance. When this goal is realized, war will never be the same.

The cognitive and autonomous levels of the latest weapons will become more capable in the years ahead. An AI-enabled kill web will increase the tempo, precision, and lethality of war by linking more sensors and shooters. Gathering, analyzing, organizing, sharing, and synchronizing this volume of time-sensitive multidomain targeting data is no longer effective at human decision-making speeds. An effective kill web has the potential to coordinate the attack of swarms of loitering munitions and UCAVs to form a persistent, precision-fire "kill box" over specified areas of the battlespace. Tactical victory can be turned into operational success if loitering munitions and UCAVs are used en masse instead of penny-packets to support combined-arms forces.

Future wars will still be fought by humans, but AI will change how wars are fought in all domains. AI promises the ability to generate exceptional speed, synchronicity, simultaneity, and accelerate the tempo of war across multiple domains. The result will be overwhelming. This is the kill web. The first to operationalize an AI-enabled kill web, even for brief periods of time, may establish fires dominance throughout the battlespace and gain a war-winning advantage. The striking power of a kill web will overmatch the enemy with furious, synchronized attacks, and at such tremendous speed, it will appear as if the opponent is standing still. Understanding the kill web, therefore, is paramount to understanding the future of warfare.

Drone Swarms have been discussed for decades but true swarms have not yet appeared in combat. In this US Navy concept drawing, drones are launched from a large transport aircraft to saturate the battlespace. (US Navy image)

CHAPTER EIGHT

The Super Swarm

> Think about it: in a congested city, teams of tiny quadrotors could buzz around to gather intelligence. Tank battalions could be overrun by miniature attack drones diving in from all directions at once. At sea, thousands of small drones could sweep in to attack a warship; many might be shot down, but others might make it through, destroying radar and leaving the ship defenseless.[1]
>
> DAVID HAMBLING, AUTHOR OF *SWARM TROOPERS, HOW SMALL DRONES WILL CONQUER THE WORLD*

Under a gray sky at Sevastopol Naval Base, the Russian Fleet's home base at the tip of Crimea, the captain of the frigate *Admiral Makarov* is standing outside on the port side of the ship, smoking a cigarette. The weather is cool with a fine drizzle. He turns up his collar as a cold gust of wind blows against the ship.

As he smokes his cigarette, he thinks about the last few months of war. His ship has been in combat since the invasion on February 24, 2022, and has launched dozens of missiles against Ukrainian targets. He has also cautiously avoided the dreaded shore-launched Ukrainian Neptune missiles by keeping away from the coast. Other Russian ships have not been as lucky, as the previous flagship of the Black Sea Fleet, the cruiser *Moskva*, was sunk when the Ukrainians ambushed it with missiles and a TB2 drone. Since *Moskva*'s sinking in April, his ship has had the honor of serving as flagship.

"With great honor comes responsibility," he mutters out loud to himself. He worries the Ukrainians will try to strike his ship to gain another propaganda victory.

He also thinks he worries too much. Sevastopol is a protected harbor, heavily defended by Russian air-defense systems and out of the range of Ukrainian weapons. The Ukrainians have no navy. The only threat is from missiles and drones. Still, he never wants to underestimate the Ukrainians. The captain of the *Moskva* is proof of that assumption.

Suddenly, air-raid sirens blare from multiple locations on the shore. An explosion erupts in the sky over Sevastopol as a surface-to-air missile explodes in a ball of flame.

In a terrifying moment, he stares at the explosion for a few seconds and then scans all around. He can't see what the air-defense missiles are shooting at, but they must be Ukrainian unmanned aerial systems, possibly the feared Bayraktar TB2, but he has no way of knowing. In haste, the captain throws away his cigarette and rushes to the bridge.

He flings open the hatch to the bridge and sees several startled sailors at their stations. They all stand at attention. A young ensign, with a look of terror on his face, turns to the captain and salutes.

"Report!" the captain shouts.

"Captain, we … don't know," the ensign stammers.

"Get on the radio, you fool! Call Sevastopol Base headquarters," the captain bellows angrily. "Find out what is going on and sound all sailors to their Battle Stations!"

An Mi–8 *Hip* helicopter flies low over the ship, its rotors beating the wind. The captain sees the tail rotor of the *Hip* as the helicopter speeds eastward toward Sevastopol Bay's fishing port.

The ensign activates the alert, ordering all crewmembers to battle stations. A loud klaxon sounds on the ship. He struggles with the radio. The captain, impatient, grabs the transmitter.

"Sevastopol Base, this is *Admiral Makarov*," the captain shouts into the transmitter. "What is going on?"

"We are under drone attack," the voice on the other end of the radio answers. "Take immediate action to protect your ship."

A commander, with his shirt unbuttoned, enters the hatch of the bridge with two armed sailors. The commander salutes the captain.

"Commander, take over the helm and sail us out of the port," the captain orders.

A second explosion detonates in the sky above Sevastopol, followed by several bursts of light from inside the city. Smoke billows skyward from several locations in the naval base.

The commander drops his salute at the rushes to get the ship underway.

"Activate ship air-defense systems," the captain orders. "I will not have a blasted Ukrainian drone hit the *Makarov*!"

With the klaxon wailing the captain looks out the window. The helicopter catches his attention. He watches in fascination as the *Hip* suddenly starts firing its machine guns into the water about six hundred meters away from them.

"What the hell is that fool shooting at?" the captain asks.

No one on the bridge dares answer. All of them watch as tracer rounds shoot from the *Hip* into the water.

"Commander, anything on our radar?" the captain demands.

"No, captain. Only our helicopter and friendly ships."

"Ensign, give me your binoculars," the captain growls, turning to the hapless ensign, "and get on the radio to talk to that helicopter. Find out what the hell he is shooting at."

The ensign pulls the binoculars from his neck and hands them to the captain.

Grabbing the binoculars, the captain scans the area while the chopper hovers above a stretch of the bay. Because he cannot see a target, he can only observe machine-gun bullets striking the water and erupting into miniature geysers. He then detects a wake in the water.

"What the hell?" the captain announces out loud. "Torpedoes?"

With machine guns firing, the *Hip* flies closer to *Admiral Makarov*.

With the binoculars tight against his eyes, the captain finally sees what is coming at them. It is a small surface craft, barely discernible, and partially submerged. Probably unmanned, the captain guesses. It is very near and closing distance fast.

"Right full rudder," the captain shouts.

"Right full rudder," the commander at the helm repeats.

The ship's engines surge but the move is too late. The Ukrainian unmanned surface vessel (USV) slams into the ship's starboard side and explodes. The captain holds his breath.

There is smoke and flames and then nothing. The USV struck the ship, but *Admiral Makarov* is still sailing. The captain breathes again. "Ease your rudder. Damage report!"

***

Only *Admiral Makarov*'s captain and the Russian crew are aware of the specifics of what transpired on the bridge and what was spoken on October 29, 2022, during the Ukrainian assault on Sevastopol. Ukraine launched a multidomain unmanned systems attack on the Russian Black Sea fleet. The assault was filmed in stunning real-time video by Ukrainian USV sea drones, which was then uploaded to social media hours later. Unmanned air and maritime systems were used in the Ukrainian operation. In order to create a plausible diversion, Ukrainian UASs and loitering munitions targeted the port infrastructure. Meanwhile, a number of USVs secretly entered Sevastopol harbor and attacked the frigate and several other ships. During the attack, *Makarov* was slightly damaged, with one of its radar systems disabled, but the ship continued to sail. USVs also hit two other Russian vessels. The Russian command was terrified by this raid on the well-defended naval port, which led the Russian Navy to withdraw many of its ships east, away from the Black Sea to the more secure Sea of Azov.[2] In this amazingly well-coordinated operation, Ukraine conducted the first combined air–sea unmanned operation in the Russia–Ukraine War. It is easy to envision robotic platforms being widely used in comparable operations as they become smaller, more accessible, capable, and networked.

Swarming as a Tactic

Swarming is "engaging an adversary from all directions simultaneously, either with fire or in force."[3] Swarming is not new and has long been a tactic in war. In a

Robotic systems, such as unmanned aerial systems and robotic ground systems, are the new weapons of war. This image, where one soldier commands drones and robotic armored vehicles, is a visualization of future warfare by the US Army Combat Capabilities Development Command. (Jamie Lear, US Army Combat Capabilities Development Command C5ISR Center)

masterful study on swarm tactics, Sean J. A. Edwards listed 23 swarming case studies, "ranging from Scythian horse archers in the fourth century BC to Iraqi and Syrian paramilitaries in Baghdad in 2003, to understand swarm tactics and formations, the importance of pulsing, and the general characteristics of past swarms."[4] An example of swarming Edwards did not mention occurred when Lt Col. George Armstrong Custer and a portion of his 7th Cavalry Regiment was swarmed by the Lakota Sioux, Northern Cheyenne, and Arapaho tribes at the Battle of Little Big Horn in 1876. Custer blundered into one of the largest Indian camps ever assembled in North America, tried to fight his way out, and was overwhelmed by Indian fighters. The complete battle narrative is much more nuanced than this but, for our discussion of swarming as a tactic, Custer's Last Stand offers a vivid visualization of how a swarm can win. In the end, Custer was killed along with 268 other troopers of the 7th Cavalry. The Indians, whose total numbers may have been as high as 2,500, lost up to about three hundred men, but won their greatest battlefield victory over the US. As Edwards explained: "Swarming occurs when several units conduct a convergent attack on a target from multiple axes. Attacks can be long-range fires or close-range fires and hit-and-run attacks. Swarming can be pre-planned or opportunistic. Swarming usually involves pulsing where units converge rapidly on a target, attack and then re-disperse."[5] This is exactly what the

Indians did to Custer and his 7th Cavalry. The lesson here—do not let the enemy swarm your force; do not be Custer.

On a clear night with a full moon shining, sometime before 4 am on September 14, 2019, 20 delta-wing Iranian unmanned aerial vehicles (UAV) fanned out in a wide formation, flying fast over the empty desert terrain. The UAVs flew low to avoid Saudi radar detection and only gained altitude in the very last minutes of their flight. As they neared the objective of their attack, the Aramco Abqaiq oil facility, they climbed and then dived, kamikaze-like, into the most critical oil facilities, striking their targets with pinpoint accuracy. At nearly the same time, four cruise missiles hit the remote oil complex at Khurais, 212 kilometers (132 miles) to the southwest. Explosions rocked the ground as the sky above Abqaiq and Khurais lit up with fire. As the conflagrations raged, Riyadh's oil production was temporarily cut in half and oil prices soared 20 percent. The Iranians coupled their attack with a disinformation campaign that the weapons used in the raid came from their Houthi proxy forces in Yemen, but the wreckage of the suicide drones salvaged once the fires were extinguished were identified as "Iranian Delta Wing UAVs" by Saudi Ministry of Defense officials. Saudi Arabia had just experienced what Israeli

A small unmanned aerial system hovers during an exercise at Joint Base Lewis-McChord, Washington, on August 10, 2020, as part of the OFFensive Swarm-Enabled Tactics (OFFSET) Program. Through OFFSET, the US Navy is partnering with Defense Advanced Research Projects Agency to develop software, physical autonomous systems integration, and test and evaluation support for autonomous swarming tactics. (US Navy photo)

air- and missile-defense expert Uzi Rubin called "a kind of 'Pearl Harbor' attack" using a swarm of low-altitude loitering munitions and cruise missiles.

The attack on the Saudi oil complexes highlighted how drones have become effective long-range precision-fire systems. During the Second Nagorno-Karabakh War, the Azerbaijanis demonstrated the use of loitering munitions and Bayraktar TB2 unmanned combat aerial vehicles (UCAVs) in groups of four to twelve systems. Using UCAVs and Israeli-made loitering munitions, Azerbaijan ruled the skies early in the war, eliminating Armenian air defense, electronic-warfare systems, and command and control. Since the Russian invasion of Ukraine on February 24, 2022, both sides have learned the value of drones of all types. Today, the proliferation of cheap and inexpensive UAS technology provides nearly any force a hip-pocket air force and long-range precision fires at its disposal.

Although swarming tactics are being tested by the Chinese People's Liberation Army (PLA) and the Russians, swarms of hundreds of UAVs have not yet been observed in combat. In 2018, the China Electronics Technology Group Corporation and Tsinghua University released a video of a swarm zooming in improvised, network-generated flight patterns. In October 2019, the PLA put on a display of its latest, high-end UAS during the Nation Day Parade. The PLA is also practicing with massed UAS swarms. As UASs gain more range and artificial-intelligence-driven autonomy, their ability to think independently and operate in swarms of hundreds of UAVs will create a significant threat.

> Although it may sound like a page out a science fiction novel, the only thing that could probably counter such a dense swarming attack on ground forces or a garrisoned force would be for those forces to have their own counter-swarm swarms at the ready. This would result in dozens or even hundreds of mini kamikaze dogfights in the sky—a life and death suicide struggle among diminutive hive-minded flying robots.[6]

Just as air defense was crucial to the success of combined operations in the 20th century, counter-UAS operations and smashing the swarm are vital to surviving and winning in today's battlespace.

During the unmanned systems attack on Sevastopol, the Ukrainians used just enough air and sea unmanned systems to overcome the Russian defenses to achieve their goals and, using unmanned systems, never put a Ukrainian soldier in jeopardy. With warheads too small to sink a frigate, the aim of the attack was merely to cause as much damage as possible, secure a propaganda victory, and force the Russians to realize Sevastopol was no longer a sanctuary for their fleet. The Ukrainians accomplished these goals, but this was not swarming. It was the employment of separate teams of robotic systems.

In the significant wars of the early 2020s—Second Nagorno-Karabakh, Israel–Hamas, and Russia–Ukraine—unmanned systems have been employed piecemeal, or in small packets, but not in swarms. In the Sevastopol attack, the individual systems operated under human control as individual weapons. Learning from these wars, we should consider a high-low mix of systems for military force structure.

> Such a military would be composed not of small quantities of large, exquisite, expensive things, but rather large quantities of smaller, lower-cost, more autonomous, and consumable things and, most importantly, the digital means of integrating them. These kinds of alternative capabilities exist now or could be rapidly matured and fielded, in massive quantities, within the window of maximum danger.[7]

It is also important to differentiate between the tactic of swarming and the concept of an artificial-intelligence-enabled swarm.

Definition of the Super Swarm

In the classic western *The Wild Bunch*, the movie starts with a group of children watching a fight between a scorpion and a swarm of ants. The scorpion is in the middle of an ant mound and is doing its best to fight off scores of ants. As the ants continue their relentless attack, the scorpion is losing the fight for its life. Ants attack the scorpion from every direction. The ants are "smart" and act as individual intelligent agents; they communicate with each other, they are disposable and sacrifice themselves; they have a purpose; they move with the swarm, and they keep attacking the scorpion until it drops and lies motionless.

In a more recent film, the 2019 action movie *Angel Has Fallen* dramatically shows the power and ferocity of an artificial-intelligence-enabled drone swarm. In the swarm scene, hundreds of fast, small loitering munitions attack the US president's Secret Service security detail. The drones move at the speed of racing drones. At first, they appear flying low above the trees and the president's security detachment cannot identify what they are, thinking they might be a flock of birds. As the drones get closer, they attack with precision and dive into the agents. The carnage is precise. Exploding upon contact, they kill everyone except the hero and the president, who dive into the water and use it as protection. The ferocity and speed of this scene is stunning. The hero Mike Banning (Gerard Butler) barely saves President Trumbull (Morgan Freeman), who is the target of the drone swarm attack. Knowing they are powerless against these drones, one can only imagine the sense of hopelessness these Secret Service members would experience in a real-world scenario. Their years of training, education, and superior small-weapons skills in this situation are useless against the drone swarm.

Both the ants in *The Wild Bunch* and the drone swarm in *Angel Has Fallen* represent artificial-intelligence-enabled drone swarms. Every ant in the swarm is an intelligent agent, with sensor-and-strike ability—communicating with each other and attacking autonomously—and are commanded by one purpose, the protection of the ant queen. In *Angel Has Fallen,* the drones overwhelm the agents with explosive power.

Each drone is an intelligent agent, with sensor-and strike-ability—communicating with the other drones in the swarm and attacking autonomously—and are commanded by one purpose, killing the president's security detail. Spoiler alert: the bad guy in charge of the drones spares Mike Banning and the president in order to advance them to the next plot point.

Hollywood is not reality but, in both instances, these visuals give us an idea of what a drone swarm with artificial-intelligence (AI) capabilities does. To differentiate it from drones that are human-controlled and use swarm tactics, an AI-enabled swarm will be referred to as a "super swarm" for the purposes of this discussion. A super swarm is a group of AI-enabled, disposable, autonomous robotic systems that cooperate and are directed by a single "pilot" to attack multiple targets simultaneously from different angles. The robotic systems in the swarm act as "intelligent agents" of a collective, performing actions to achieve goals set by the AI. A super swarm combines the tactic of swarming (convergent, multiple, and relentless attack) with a network of intelligent agents. To accomplish this, AI organizes, navigates, synchronizes, and directs the super swarm. One human "pilot," or the AI, steers the swarm and regulates its activation or deactivation. Let's break down this definition to explain how the super swarm will work by creating a simple word equation.

SUPER SWARM WORD EQUATION

SUPER SWARM = AI-ENABLED + LARGE NUMBERS OF DISPOSABLE AUTONOMOUS ROBOTIC SYSTEMS + SWARM TACTICS

First, AI must enable the drone swarm to operate in unison as a group and move in the direction established by a human "pilot" or autonomously with pre-programmed instructions. The US Navy already proved this capability in 2017 during LOCUST (Low-Cost UAV Swarming Technology) tests.[8] The LOCUST program involved AI-enabled UAVs, including Raytheon's Coyote loitering munitions,[9] in a network to collaborate, share information, sense, strike, and assess. In hypothetical war scenarios, super swarms have also been used to repel a Chinese invasion of Taiwan:

> Wargames that the US Air Force has conducted itself and in conjunction with independent organizations continue to show the immense value offered by swarms of relatively low-cost networked drones with high degrees of autonomy. In particular, simulations have shown them to be decisive factors in the scenarios regarding the defense of the island of Taiwan against a Chinese invasion.[10]

On March 16, 2021, the US Department of Defense published a report that the autonomous-swarm/strike-loitering munitions effort will use Raytheon-made Coyote Block 3 drones launched from ships.[11]

Next, each drone in the swarm must be smart enough to communicate with every other system in the swarm and be cheap enough to sacrifice to neutralize the target. For instance, a combat swarm might consist of 50 Coyote loitering munitions with the mission of destroying a dozen critical air-defense systems in a portion of the battlespace. Each system in the swarm will sense its surroundings, understand PNT (positioning, navigation, and timing) to accurately and precisely determine

During the Second Nagorno-Karabakh War, the Azerbaijani military used loitering munitions, unmanned combat aerial vehicles, and long-range precision fires to locate, disrupt, destroy, and defeat the Armenian defenders in a rapid 44-day campaign. This was the first war in history where robotic forces played a decisive role in securing a military victory. (Screenshot from an Azerbaijani MOD video)

location, communicate with each member of the swarm, and fly with the swarm toward the target. As soon as the target is identified, AI will decide which systems in the swarm are best suited to attack the target. "A Swarm of armed drones is like a flying minefield. The individual elements may not be that dangerous, but they are so numerous that they are impossible to defeat ... Minefields on land may be avoided; the flying minefield goes anywhere."[12] Multiple systems will attack simultaneously from multiple directions to overwhelm enemy systems. A new system replaces the one destroyed if it cannot complete its attack. All systems will move on to the next target as soon as the target is determined to be neutralized.

Last, the group acting in unison must execute swarm tactics, where individual intelligent agents conduct convergent, multiple, and relentless attacks:

> Being networked together ... these swarms will be extremely hard to defend against using even the best SHORAD [short-range air defense] systems in development today. It's the saturation nature of the attack, the size of the attackers, and the fact that they work as a coordinated swarm, employing dynamic tactics to see as many in their company survive long enough to make their suicidal attack, that makes them so deadly ... Just the knowledge that such an attack is possible would be psychologically stressful and demoralizing for troops on the ground.[13]

Super swarms, which are made up of hundreds of drones and connected as sensors for other long-range precision fires, have the power to produce decisive outcomes and generate battleshock. When that happens, combat will never be the same.

CHAPTER NINE

Visualize the Battlespace

> Clausewitz had a term called coup d'oeil, (a great commander's intuitive grasp of opportunity and danger on the battlefield). Learning machines are going to give more and more commanders coup d'oeil.[1]
> ROBERT WORK, FORMER DEPUTY SECRETARY OF DEFENSE

The weather is cold. It is BMNT (Before Morning Nautical Twilight) and the sky is graying. The sun will be up soon. A battle looms and its prospects appear as gray as the dawn.

"Foresight is hard … Harder still when the enemy is winning," Daniel says as he surveys the ground where he expects to fight.

"Sir?" a young lieutenant, standing by his commander's side, asks.

"Nothing," Daniel replies. "Get the men into position. We'll be in a fight soon."

"Immediately, sir," the lieutenant answers and then rushes off to issue the orders.

Daniel knows his men will soon be in an uneven battle, fighting for their lives. He can see it all in his mind's eye. An open, rolling woodland of first-growth pines and hardwood trees border a wide-open pasture. There is a rise at the center of the pasture, then a general downward slope to the northeast to a wide river. The rise offers him a defensive advantage. It's tall enough to hide his men from the enemy, but only until his opponents reach the crest of the rise. Behind this reverse slope, the river is a problem, as the waterway is too deep and swift for his men to ford if they carry their weapons and equipment. Outnumbered and pinned by the river, his troops must either win here or die.

Daniel thinks through the enemy's potential moves and countermoves. He lays out various courses of action in his mind with moves and countermoves. How will I use this ground to our advantage? How will my soldiers win this day?

He reflects on the most recent intelligence assessments. The enemy is only a few kilometers away. Intelligence, surveillance, and reconnaissance information indicates his adversary is advancing swiftly, ready for the kill. He knows that, within the hour, the enemy will arrive at his location in strength. He has one desperate chance left.

At 6:45 am, as the sun rises, Daniel sees a forward detachment of the enemy emerge from a clump of trees to the southeast. The fight is about to begin. Imagining what the enemy commander is thinking, he is confident his opponent is thrilled with the prospect of bagging his prey. The enemy commander will see the Americans occupying positions on the forward slope of the rise; with the river at their backs, there is little potential for escape.

Daniel receives more reports as his troops observe the enemy. The foe is still out of range. Yesterday, Daniel established an engagement area (EA), issued orders, and rehearsed his troops. He established two lines of defense, with one unit on the forward slope and a second unit on the reverse slope. The forward slope unit is a delaying force. The unit on the reverse slope is the main effort. He also designated a mobile unit to be his reserve and positioned them to the left of the second line. The subordinate unit leaders of these formations know to wait for his signal to fire. Timing is important.

Seconds pass. He waits as the enemy moves more troops into the EA. After a few more moments, Daniel gives the order: "Commence firing!"

The unit on the forward slope engages the enemy. Artillery fires on the advancing foe as the enemy rushes the first line. Rifles blaze. Enemy soldiers are hit. Smoke masks the American positions, but Daniel can see the enemy is still advancing, seemingly sure of victory.

Now, enemy artillery gets into the fight and blasts Daniel's troops on the forward slope. The enemy concentrates his artillery fire on these defenders. The fire is intense. Daniel worries as he sees his forward forces waiver. Soon some Americans start falling back behind the hill toward the second line. Smoke rolls across the battlespace.

Daniel knows the enemy can only see his forward positions. He understands the battle is reaching its critical point. He guesses the enemy commander must sense victory as he observes a separate enemy mobile force race forward to hit his left flank. Patiently, Daniel watches and waits as the enemy closes the distance on his left. His small mobile reserve masked behind the hill is his trump card and he hopes it will be enough. As this antagonist moves into striking range on his flank, Daniel orders his reserve to counterattack the enemy mobile unit. His counterattack force blasts away at the unsuspecting enemy. Surprised, the enemy is caught in the open with no cover. Many of the adversaries fall and the rest disengage in confusion.

Daniel watches. He knows the next few minutes will tell who wins or loses this battle.

The enemy, undeterred by the repulse of his mobile unit, commits his remaining force at Daniel's weakened defenses on the forward slope of the hill. Dense clouds of gray smoke sweep across the field. Finally, the American line breaks and appear to be in full retreat, running over the crest of the hill and toward the river. The enemy senses victory and charges, seizing the crest.

This story is not about a battle in eastern Europe in modern times. It describes the battle of Cowpens, South Carolina, during the American Revolutionary War, on January 17, 1781.

Brigadier General Daniel Morgan watches the fight unfold with a keen sense of satisfaction. His second line of Continental regulars is waiting on the reverse slope of the hill, ready to fire. The British race down the far side of the hill but suddenly stop. In shock, they realize a thick line of blue-clad American regulars is standing at the bottom of the hill, in front of them, with muskets ready, and waiting for the order to fire.

Morgan knows this is the decisive moment. He shouts the order to his men: "Fire!"

A thousand muskets fire as one and the noise from the volley shakes the heavens. A fierce fusillade of bullets tear into the British mass. Redcoats fall in heaps as others stagger or fall to their knees, wounded. The surviving British are only 30 paces away and cannot return fire as they have not had a chance to reload their one-shot muskets. For a moment there is an awful silence.

"Charge bayonets!" Morgan orders.

"Bayonets! Charge!" The order is relayed by Morgan's subordinate commanders.

The Americans cheer and charge with gleaming bayonets, ready to finish the fight with cold steel. The sight unnerves the British, and the Redcoats turn and flee, running for their lives. In a moment, the British have lost all sense of order and it is every man for himself. Lieutenant Colonel Banastre Tarleton, the British commander, curses and flails at the air with his sword, but cannot stop the rout of his forces. He barely escapes with his life.

The American militia, who manned Morgan's first line, have re-formed behind the regulars and are now running forward on both flanks to pick off the fleeing Redcoats with their long-barreled Pennsylvania rifles. Morgan's men envelop the retreating British from the front and both flanks as American cavalry charges the British right flank and rear. Tarleton's Redcoat formations disintegrate, dropping their muskets and abandoning their wounded; their only thought is to get away from the relentless Americans.

Daniel Morgan's brilliant action at the Battle of Cowpens is a stunning masterpiece of leadership, visualization, and foresight. Morgan's Americans suffered 25 killed and 124 wounded. The British lost 110 killed, 229 wounded, and 629 captured. Surviving British forces fled the battlefield, leaving two cannons behind. Cowpens was one of the most important victories of the American Revolution, as the results of the battle and the subsequent engagement at Guilford Courthouse, on March 15, 1781, forced British Gen. Charles Cornwallis, the overall commander of the British forces in the southern theater of war, to withdraw to Yorktown, Virginia. At Yorktown, Cornwallis was surrounded and besieged by a combined American–French force under Generals George Washington and Jean-Baptiste-Donatien de Vimeur, comte de Rochambeau. Cornwallis surrenders on October 19, and the Battle of Yorktown becomes the decisive engagement of the war for American independence.[2]

Daniel Morgan was an exceptional commander. He had the ability to see a battle play out in his mind's eye and then test the success or failure of each move before

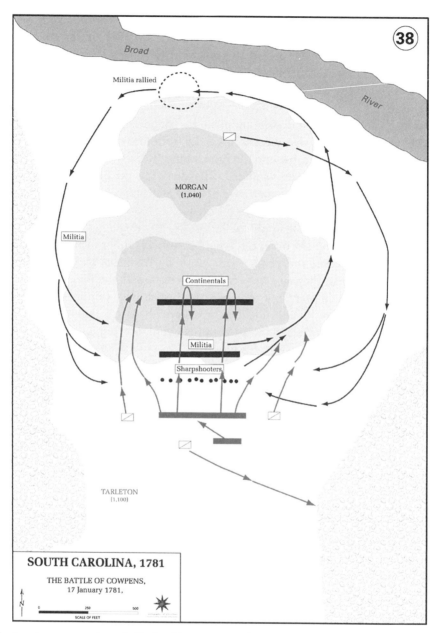

The American commander at the Battle of Cowpens was gifted with the ability to visualize the battlespace and arrange the events of the battle in his mind's eye. The *coup d'oeil* demonstrated by Gen. Daniel Morgan is difficult to replicate in today's hyper-fast, complex, multidomain battlespace. Technology must assist in the effort. The Battle of Cowpens—January 17, 1781—during the American Revolutionary War was a turning point and a rare example of a double envelopment. The battle was a decisive victory for the Americans and lasted less than an hour. (US Army photo)

it actually occurred. This remarkable foresight enabled him to achieve something that rarely happens in the story of warfare: A double envelopment and a crushing, decisive tactical victory. He had the gift of *coup d'oeil*, which means the stroke of the eye, what Clausewitz would later describe as the "ability to see things simply, to identify the whole business of war completely with himself, that is the essence of good generalship. Only if the mind works in this comprehensive fashion can it achieve the freedom it needs to dominate events and not be dominated by them."[3] Today, however, war is much more complex than it was in 1781. Unable to see the multidomain battlespace with human senses alone, today's commanders will only gain *coup d'oeil* through experience in constructive, virtual, and live simulations, and will require technology to enable them to visualize the multidomain fight.

Maneuvering across multiple domains is nothing new. Commanders have done this across land and sea in ancient times. In the 20th century, the air domain became vital. Multidomain operations across the five domains (land, sea, air, cyber, and space), due to new technologies and capabilities, are changing the methods of war. Advances in computers, space operations, artificial intelligence (AI), microminiaturization of electrical components, and robotics are driving these changes. As systems and weapons become smarter and are connected, an Internet of Battlespace Things, similar to the commercial Internet of Things, is emerging. War will be faster, moving at the speed of electrons, and deadlier than ever before. The fusion of networked sensors with projectiles, and the synchronization of the kill web by AI, will strike targets in multiple domains at hyper-speeds. To act decisively in this complex environment, commanders must visualize the battlespace, see actions as they occur, and predict possible effects more rapidly than the enemy.

As with Daniel Morgan at the Battle of Cowpens, human decision making rests primarily on pattern recognition. Morgan knew the patterns of war of his time and applied counter-patterns he thought would win. Commanders observe, orient, decide, and act (the OODA loop) by recognizing the pattern the enemy has presented and rapidly applying a counter-pattern. If you can do this faster than your opponent, the enemy will appear to be moving in slow motion. It is as if a boxer is landing four or five blows for every one blow of their adversary. Accelerating the OODA loop will require enhanced cognitive computer systems that can depict the multidomain battlespace in actual time on an All-Domain Common Operational Picture (ADCOP).[4] An ADCOP enables a commander to see appropriate relevant information, especially friendly and known enemy actions, in all five domains (land, sea, air, space, and cyber). This ADCOP is shared by all pertinent commands in near real-time.[5]

In business, decision makers use versions of COPs to visualize complicated information in several domains. Real-time stock analysis software depicts a common picture for multiple users to understand and act on trading data. Military developers can learn from these private industry best practices. Although the consequences are less lethal in commerce, the outcomes often involve the success or failure of multi-million-dollar efforts. Commercial software systems, like their military counterparts, use a user interface (UI) as the point of human–computer interaction and communication with a device. If designed properly, UI speeds up synergy and comprehension. The commercial electronic and gaming industries focus on clear, easy-to-use, and self-evident UI. The best UI solves the problem without overwhelming the user and makes operations easier and thus more effective. Robotic systems can now be controlled by the human mind using a brain–machine interface (BMI) connection. In February 2023, the Australian Army released a video that demonstrated how a human using a BMI and a Microsoft HoloLens can control a robot with his mind, without voice or manual controls.[6]

The need for an ADCOP is urgent. Military information needed for cross-domain maneuver is neither holistically displayed in all five domains, generated automatically, nor easy to comprehend. It is usually depicted in large headquarters on multiple

An example of an equipment suite to display a multidomain Common Operational Picture is provided by Elbit Systems' Torch-X C4ISR solutions. With seamless integration of sensors, effectors, and communication systems, and full support for manned and unmanned autonomous platforms, the advanced systems offer enhanced cross-force coordination, strategic planning capabilities, comprehensive battle management, tactical operations, survivability, and lethality. Torch-X C4ISR solutions feature AI-based decision-support tools to reduce cognitive load at all echelons, facilitating optimal decision making and planning processes. (Elbit Systems)

screens cluttered with icons in layers of increasing complexity. Staff personnel working within these information silos create most of the data that appears on these screens, with the corresponding human time lag. This data is assembled into information manifested as a "kill chain" to attack targets. The traditional kill chain (find, fix, track, target, engage, and assess) works fine for single actions, but will take too long to apply in multidomain operations against a peer adversary. A solution is to learn from commercial, real-time systems and information presentation in real-time video games, to develop an ADCOP that will assist in executing an AI-assisted kill web. Taking the best lessons from commercial UI design, here are seven principles to develop an ADCOP.

Holistic View

In any battle, the whole is more than the sum of its parts. Actions occurring on the flank of any battlespace can quickly turn decisive. A commander must see all the factors of the fight as they happen in the battlespace in as close to real-time as possible and be able to use this information to issue orders, make plans, and develop future operations. A holistic view combines several domains and the actual world to see the complete picture. Cyberwar actions, integrated with the actual world and other domains, are superior to a single domain visual. Seeing friendly and enemy cyber actions as they occur, overlayed onto a real-world setting, allows the commander to form a holistic vision of the battlespace. In the commercial sector, the Johns Hopkins COVID-19 dashboard is an example of a holistic COP.[7] From this dashboard, the user can track the number of global cases and then dig deeper to determine the figures by nation, city, and region. This dashboard was so up to date and trusted that many health providers and governments consider it the standard for COVID-19 information. The UI depicts the data on a world map and the user can tap down to any desired area. In addition, the dashboard depicts the testing, tracking, contact tracing, information on vaccines, news, and resources, providing a holistic view of the current pandemic situation. Similarly, a military ADCOP must depict the actions and effects of multiple domains over an area of interest.

Self-Evident

Anything displayed on a COP should be axiomatic. Symbology that is not intuitive fails. Consider the smartphone. Users purchase a phone and begin using the device with no training because its iconography is self-evident. The touch-and-activate UI that then cascades into expanding information on demand is what should be the standard for an ADCOP. Real-Time Strategy (RTS) computer and console games provide an example of self-evident displays that take little training time. In these games, the UI depicts multiple actions in real-time. Timing is critical in massively multiplayer online (MMO) games where players battle other human players in sports or combat genres. A generation of people have been playing MMO games since the beginning of the internet. Games such as "World of Tanks" and "War Thunder," which boast thousands of users every day, are realistic and immersive experiences. World of Tanks, by a Belarusian company Wargaming.net, claims 180+ million players worldwide.[8] In these games, speed matters as players operate tanks and other military equipment in multiplayer battles over realistic terrain in real-time. The simple and intuitive UI best practices of these games are the principal reason for military developers to take notice.

Minimalist

The best commercial UI designs offer only the required or essential features. The goal of every UI design should be simplicity and ease of use. A classic video game titled "Metroid Prime," developed by Retro Studios, is an excellent example of a minimalist UI design. The UI in Metroid depicts everything in the game in a diegetic[9] view from the character's helmet heads-up display (HUD). In this diegetic view, we see the elements in the game in a first-person perspective. The HUD in these games is usually a piece of equipment, such as the helmet visor from a powered suit of armor. In games with vehicles, we see a virtual cockpit. Military designers can learn from this diegetic view for helmet-mounted or virtual-cockpit systems. A helmet-mounted ADCOP, similar to the helmet capabilities of an F-35 pilot, which provides an augmented-reality view of the real world, could benefit from the minimalist design concept for a heads-up display from video games like Metroid.[10]

Reduce Cognitive Load

Data overload is a hazard. In battle, data overload is deadly. An ADCOP, that translates mountains of data into useable information, with recognizable symbology and other cues, can reduce the cognitive load on the warfighter similar to the navigation software on your smartphone. Today, most people use their smartphones to navigate. The navigation system tells you point-by-point directions by voice and indicates the vehicle direction and location on a map in real-time. No need to glance at notes or a paper map. The driver remains focused on driving while receiving information. This is an excellent example of a system that speeds up the driver's OODA loop while reducing their cognitive load.

Multiple Forms of Feedback

A smartphone helps a driver navigate their car from one point to another with audio and visual cues. For an ADCOP, designers must consider multiple forms of input, including tactile stimulation such as the vibration stimulator in an Xbox controller.

With this additional UI, the controller vibrates when triggered by explosions or dangerous events in the game, alerting the player. In a similar fashion, the UI of an ADCOP might alert the operator to a dangerous or significant event.

Configurable and Locus of Control

The UI should be customizable and configurable so the operator can fit it to the circumstance, such as adding shortcuts to display or executing repetitive actions. In addition to being configurable, the UI of the ADCOP must give the operator confidence they control the system, not the other way around. The ADCOP is a primary means for human-on-the-loop (HOTL) control of AI. User control, called "Locus of Control" in psychology, is as vital in an ADCOP as it is in the rifleman's confidence when looking though a rifle sight, squeezing the trigger, and firing a shot that knocks down a target. They gain this confidence when actions deliver multidomain effects in real-time in the battlespace.

Automated

To conduct cross-domain maneuver in real-time, we must act faster than the current human staff process allows. Staff silos must give way to secure and automatic reporting and integration, with HOTL oversight. Fighting in real-time requires an ADCOP updated by Internet of Battlespace Things sensors. An example of a user-friendly, commercial real-time, automated COP is Meta-Trader 5 (MT5), developed by MetaQuotes Software Corporation. Foreign exchange (forex) traders use MT5 to get advice, news, alerts, and analysis to trade forex in real-time, with instant execution, from a personal computer, tablet, or smartphone. Although this software deals primarily with one domain (forex trading), it does so in real-time, with automated inputs that provide the essential information and predictive analysis for decision making.[11] Imagine the reaction of a commander with an ADCOP that is suddenly "lit up" by cyberattack alerts. To conduct cross-domain maneuver, the cyber slice of this multidomain COP must automatically depict when and where every system in a unit is operating, emitting, which is being attacked, and offer options to handle the threats.

Developing an ADCOP

War is neither a video game nor trading on the foreign exchange market, however the commercial sector does hold valuable lessons for military developers. As mentioned earlier, an ADCOP is an urgent need. To conduct cross-domain maneuver, operators must see the relevant domains on-demand, in real-time, and with predictive analysis. The predictive-analysis aspect is crucial. Real-time is not enough. The hyper-speed of future battle renders real-time events as past actions for human cognition. To adapt to changing circumstances, the commander needs enhanced cognitive AI to predict what is likely to happen next. As the real-time fight is occurring, the ADCOP must help the commander visualize the fight in "predictive time" and see how their actions might play out. Visualizing all domains in one integrated COP, and providing predictive analysis, is beyond human cognition and requires enhanced systems with robust AI. Those who master advanced methods to synchronize, visualize, predict, and execute combat in this multidomain battlespace with an ADCOP will gain a tremendous advantage.

Speed and AI

Speed is the crux of battle. Alexander Suvorov, one of the greatest commanders in Russian military history and of the era of early modern warfare, is attributed to have said, "Winning time is winning battle."[12] When the time to execute long-range precision-fire engagements is measured in minutes or seconds, this concept is more relevant today than ever before. The side that can decide and act faster than the enemy has a significant advantage. Imagine a single network that connects joint sensors in all domains—land, sea, air, space, and cyber—and includes relevant information concerning information warfare while monitoring the electromagnetic spectrum. Now picture a setting where warfighters at all levels have access to the information pertaining to their mission and can visualize the operations and intent two echelons above their own. Such connectivity speeds up decision making and expedites the kill chain. This is the promise of the US Department of Defense's (DOD) Joint All Domain Command and Control (JADC2) concept.

JADC2 is "DOD's … concept to connect sensors from all of the military services—Air Force, Army, Marine Corps, Navy, and Space Force—into a single network."[13] JADC2 promises to enable cross-domain maneuver at machine speeds. AI can assist commanders to speed up the "observe, orient, decide and act … loop" across the full spectrum of combat operations. Uber, the ride-sharing company, is used as a metaphor for JADC2:

> Uber combines two different apps—one for riders and a second for drivers. Using the respective user's position, the Uber algorithm determines the optimal match based on distance, travel time, and passengers (among other variables). The application then seamlessly provides directions for the driver to follow, delivering the passenger to their destination. Uber relies on cellular and WiFi networks to transmit data to match riders and provide driving instructions.[14]

The concept of JADC2 is to deliver digital connectivity everywhere in the battlespace. It is more than just a faster way to connect sensors with shooters. DOD is in overall charge of the JADC2 effort, with the Joint Staff working on policies and doctrine while designating the US Air Force as the executive agent for technology development.

VISUALIZE THE BATTLESPACE • 105

	Real Time Situation Report (SITREP) Alert	
EMS Electro Magnetic Spectrum Control		**CYBER**
INFO WAR	Map depicting the All Domain Common Operational Picture (ADCOP)	**SPACE**
SITUATION 2 Echelon Up		**AIR**
SITUATION 1 Echelon Up		**SEA**
SITUATION 1 Echelon Down		**LAND**
	Real Time Logistics Alert	

A standard Operational Picture (COP) is a uniform, standard display of key data that is shared by several Commands to enable leaders to better visualize layered data. With systems networked to report information, relevant data could be shown in near real time on an automated computer interface for an All Domain Common Operational Picture (ADCOP), as seen in the example above. When a status change occurs, the information tabs' corresponding colors—Green, Amber, Red, and Black—update instantly. Any tab that is chosen extends to fill the entire screen and presents thorough details about its category. Alerts in real time are displayed at the top and bottom of the ADCOP. (Diagram by the author).

The US Army is working on JADC2 issues under its *Project Convergence* program, the Air Force with its Air Battle Management System (ABMS) program, and the Navy with *Project Overmatch*. According to a July 2021 Congressional Research Center study: "Air Force officials have argued that a JADC2 architecture would enable commanders to (1) rapidly understand the battlespace, (2) direct forces faster than the enemy, and (3) deliver synchronized combat effects across all domains."[15] As it matures, JADC2 will operationalize battlespace data and transform information into understanding.

How Will JADC2 Work?

"Technology is moving fast, but human institutions and organizations aren't keeping up," MIT Sloan professor Erik Brynjolfsson, coauthor of *The Second Machine Age: Work, Progress, and Prosperity in a Time of Brilliant Technologies*, said on February 22, 2018, at a "LinkedIn Speaker Series" talk.[16] The way to keep humans operating at faster speeds is to combine humans with computer intelligence. JADC2 requires an internet cloud and a rich sensor network in the battlespace. This cloud will enable the sharing of sensor information: intelligence, surveillance, reconnaissance, and targeting data. AI will then process the data, possibly using quantum computing, to recommend courses of action and targeting options for the human commander who will command and control using JADC2. This will merge human commanders with the speed of machines. Teaming humans with computer-assisted decision making is often called the "centaur" concept of Mission Command.

Humans require a means to visualize the complex battlespace of multidomain operations. A centaur approach to Mission Command is to use the latest technology to enable that understanding. Some may find the centaur concept unsettling, but we can also view it as an evolution of what we already do today. Tomorrow, it may be considered normal. The first human to mount a horse probably unnerved

The US Army's Integrated Visual Augmentation System uses a Microsoft HoloLens to connect technology and sensors into vehicle platforms for optimal battlefield visibility. If modified with an appropriate software package, the Microsoft HoloLens based-IVAS could be re-imagined for command and control. (US Army photo by Courtney Bacon)

those who witnessed it. Those who embraced the horse developed better hunting capabilities and mounted warfare. Enemies who were unmounted were at a distinct disadvantage in speed and striking power. Humans driving cars and airplanes are part of the man–machine interface we accept as normal today. Millions of people ask questions and get advice from devices enabled by the Lexus or Siri AI. This is a centaur approach to merge human and AI capabilities to access information to empower the human to make better, informed decisions:

> As formations increase in complexity—particularly with formations designed for Joint All-Domain Operations—controlling these forces could potentially surpass the ability of human cognition, with algorithms used to help manage these forces. The US military has stated that it intends to keep humans involved throughout the decision-making process, but as US forces introduce more artificial intelligence technologies into their decision-making apparatus, distinctions among the dimensions begin to blur.[17]

In 2016, when JADC2 was being envisioned, Paul Scharre, an expert on autonomy in weapons, and a combat veteran, wrote an article titled "Centaur Warfighting, The False Choice of Humans vs. Automation." He stated that "in many situations, human–machine teaming in engagement decisions will not only be possible but preferable … The best systems will combine human and machine intelligence to create hybrid cognitive architectures that leverage the advantages of each."[18] JADC2 is supposed to be the evolution of that idea.

Between December 2019 and July 2020, the US military exercised multiple JADC2 elements. The first simulation involved the US Air Force's ABMS using joint systems from all services, including commercial space sensors. A combined wartime simulation set in the Black Sea was the focus of the second exercise that involved joint US forces operating with eight other NATO members. The focus of both tests was exercising JADC2 data sharing in simulated target engagements.

When it comes to developing any joint system, especially one that will connect sensors from all of the military services—Air Force, Army, Marine Corps, Navy, and Space Force—into a single network, it is very expensive. The Air Force version of JADC2, the ABMS, is a case in point. "DOD requested $302.3 million for ABMS in FY2021 but was appropriated $158.7 million (a $143.6 million decrease) due to unjustified growth and forward financing … DOD also requested $207 million for 5G Congested/Contested spectrum research and development, seeking to develop spectrum sharing technologies and network security architectures."[19] JADC2 clearly is a priority for DOD and will be a major enabler of centaur Mission Command. On October 3, 2022, US Army Lt. Gen. Xavier Brunson, the commanding general of the Army's First Corps, explained the challenges that lie ahead:

> This is now a whole new world, this is not what we've been doing for the last 20 years where we have gotten used to having some very robust, resilient capabilities in our brigade combat teams. A large-scale combat operation means that we now have to have complete integration of data in a distributed way, all the way from the foxhole to the fort.[20]

The future of war lies in how man and machine work together. Currently, AI is not intelligent enough to single-handedly command multidomain battles, but the combination of human cognition and artificial intelligence makes the "best" system that is both intelligent and powerful. This is "Centaur Warfighting," which takes its name from a mythological creature that's half human and half horse, combining the best of both into a single combat system. The human factor in this combination is decisive. It is the X factor that makes brittle AI a better weapon to organize a host of robotic and networked systems. We cannot create an AI that can think as a human does, nor should we want to. Merging both, however, is the next step and will separate winners from losers in the next war.

On February 24, 2023, Applied Nano Materials published a scientific abstract by Shaikh Nayeem Faisal, et al., titled "Noninvasive Sensors for Brain–Machine Interfaces Based on Micropatterned Epitaxial Graphene," which demonstrated the viability of large-scale deployment of brain–machine interfaces for soldiers to command robotic systems with their minds. Soldiers would have a voice-command-free communication to operate external devices through brain waves.[21] This is no longer a story from science fiction, but an actual capability that will only improve with

Decision Dominance is the ability for a commander to sense, understand, decide, act, and assess faster and more effectively than any adversary. In this photo, an E 2C Hawkeye from the "Sun Kings" of Carrier Airborne Early Warning Squadron (VAW) 116 conducts electronic intelligence support for USS *Nimitz* in the South China Sea on January 31, 2023. (US Navy photo by Mass Communication Specialist 2nd Class Justin McTaggart)

further research and funding. The next step in multidomain manned–unmanned teaming is a further blending of the best of man and machine, a centaur approach. Western militaries, therefore, are in a historical watershed. If we do not learn from recent wars and understand the forces disrupting our traditional methods, we will not get a second chance.

Being able to decide and act faster than your opponent is the key to victory. Detailed planning and rehearsals are essential to do this. In this photo, US Army soldiers from Apache Troop, 4th Squadron, 4th Cavalry Regiment, 1st Brigade, 1st Infantry Division, prepare to conduct a zone reconnaissance mission at the National Training Center at Fort Irwin, Calif., on April 19, 2014. Learning from the lessons of recent wars, holding a rehearsal out in the open, with no top camouflage, is no longer viable as enemy drones may be recording your every move. (U.S. Army photo by Spc. Charles Probst)

CHAPTER TEN

Decision Dominance

> Decision dominance ... is the ability for a commander to sense, understand, decide, act, and assess faster and more effectively than any adversary.[1]
>
> FORMER ARMY FUTURES COMMAND CHIEF, GEN. JOHN "MIKE" MURRAY

If decision dominance is the ability to execute a faster OODA (observe, orient, decide, act) loop than the enemy, how did military commanders achieve decision dominance in the past? This was done in the mind's eye of the commander. It was developed by experience and reinforced by studying tactical vignettes, military history, and by tactical exercises without troops and terrain walks. Today, to fight a multidomain operation and conduct multidomain maneuver, more is needed. Military technology is advancing in ways the combatants of World War II could not have imagined, but lessons in the decision dominance in war from then are applicable today. One example worth studying is the Japanese invasion of Malaya in late 1941. Although technology has affected the changing methods of war, it has not altered the nature of war. War may change, but the principles remain much the same. These principles become the rules that guide commanders in planning campaigns. "Technology governs only what methods we use to achieve military decisions. Advances in weaponry actually increase the need for generals to avoid the most heavily defended and dangerous positions and to seek decisions at points where the enemy does not anticipate strikes," said military historian Bevin Alexander in his book *How Great Generals Win*.[2]

On December 8, 1941, Imperial Japanese forces launched the invasion of Malaya from their strongholds in French Indochina. The plan to conquer Malaya was the most complex military operation conducted by the Imperial Japanese Army during World War II and one of their most successful. During 70 days of brilliant fighting, the Japanese Twenty-Fifth Army, commanded by General Tomoyuki Yamashita, generated such a rapid tempo of operations that it overwhelmed the British and Australian defenders. From December 8, 1941, to February 15, 1942, the three divisions of Yamashita's Twenty-Fifth Army, advancing through dense jungle, captured the Fortress of Singapore, a British stronghold of fortifications defended by 120,000 men that had earned the title "the Gibraltar of the East."

If the Japanese had conducted a direct naval and amphibious attack against Fortress Singapore, attacking the main strength of the British defense, the British would likely have won a defensive victory. Lieutenant Colonel Masanobu Tsuji, a brilliant zealot called the "God of Operations," planned the campaign. Tsuji intended to bypass British defenses and understood the need for speed. He sought an unconventional solution that offered the greatest chance of producing a rapid tactical and operational victory. An important lesson of the Malaya Campaign is that the enemy "gets a vote." A smart, thinking enemy who does not play by the established "rules" can turn the tempo of war against you and can gain decision dominance.

The British defenders completely underestimated their enemy. Overconfidence interfered with British judgment before the war began. Believing in their fixed fortifications, the British could not imagine any invading force could conquer Fortress Singapore. Despair and panic quickly replaced arrogance when, on December 10, 1941, Japanese aircraft sank both the Royal Navy's battlecruiser HMS *Repulse* and the battleship HMS *Prince of Wales*. Catastrophes on land followed these disasters at sea as the Japanese landed forces in Malaya. The rapid tempo of the Japanese advance, with the Japanese conducting continuous, rapid, converging, and combined-arms attacks, forced the weary British defenders back from one defensive line to another. By the time the British fell back into their citadel on Singapore Island, they were so psychologically stricken they could hardly mount an effective defense and were incapable of making timely decisions. At every turn of the battle, Yamashita's decisions and speed of action dominated the battle.

The 1941–42 Malaya Campaign is an example of operational dislocation and paralysis through dominant maneuver and decision dominance, not overwhelming firepower, that forced the British into a tempo of operations they could not match. Japanese success relied on speed of movement, information about the enemy, and integrating air, land, and sea power. The speed with which the Japanese broke through and enveloped the British defenses, the willingness of the Japanese to take risks to continue the tempo of the attack, and the exploitation of gaps in the British lines by every means became the soul of the Japanese campaign. These tactics gained tremendous temporal and psychological advantages for the Japanese and set the conditions for the operational campaign. The tempo of the attack was so fast, when compared with the British, that the British could not think ahead of each cascading crisis. The daring, rapid maneuver of the Japanese land force became the handmaiden of victory and earned Yamashita the title of the "Tiger of Malaya."

Fast forward to September 27, 2020, to Nagorno-Karabakh. Armenian forces held strong defensive positions and were confident the Azerbaijanis could not move them from their "mountain fortresses." The Azerbaijanis, however, used superior training, technology, and leadership to gain decision dominance in their

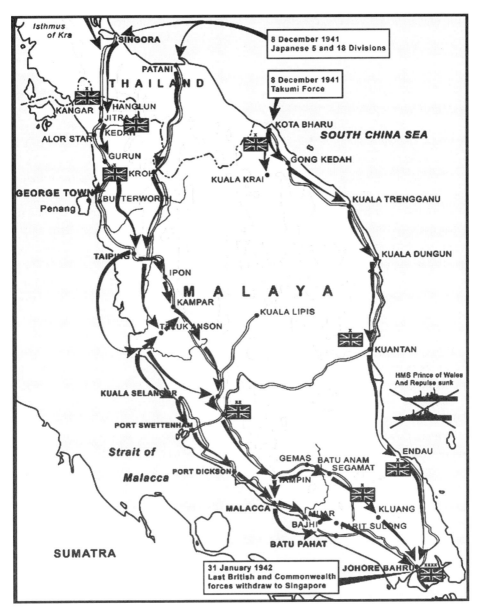

The Japanese Malaya Campaign, 1941–42, is an example of decision dominance. It was not overwhelming numbers, firepower, or technology that secured the Japanese victory, but the speed of Japanese decision making and action. They achieved a decisive victory over the British Commonwealth defenders in a brilliant campaign of only 70 days (December 8, 1941–February 15, 1942). (US Army photo)

campaign. In the years between the First and Second Nagorno-Karabakh Wars, Azerbaijan prepared. Flush with wealth from the oilfields around Baku, it built up its military forces. Strongly supported by Turkey, the Azerbaijanis purchased the latest military technology from Turkey and Israel. Even though Azerbaijan is a Muslim nation, it has a very close relationship with Israel and the Israelis were happy to sell the Azerbaijanis their latest weaponry. The way the Azerbaijanis used the systems they purchased from Turkey and Israel overturned Armenia's defensive advantages.

First, the Azerbaijanis dominated the opening phase of the war by conducting a first strike that knocked out Armenian air defense, electronic warfare, and command and control. Employing their loitering munitions, unmanned combat aerial vehicles (UCAV), and missiles, the Azerbaijanis dominated the battlespace and maintained this dominance throughout the 44-day war. Frozen in their trenches, the Armenians consistently found themselves several moves behind their opponent and appeared to be standing still as the Azerbaijani military dominated each action. Armenian troops were shackled by the persistent fear, real or imagined, of an attack from above by loitering munitions and UCAVs that could come at any time and any place. There were no major tank battles, even though both sides had sizable armor and mechanized forces.

Azerbaijan also used the streaming videos captured by their anti-tank guided missiles, loitering munitions and UCAVs to influence Armenian decision making. Every night the Azerbaijanis would barrage social media in Armenia with images of the death of Armenian soldiers and the destruction of their command posts, tanks, artillery, and vehicles. These streaming videos were usually put to music and helped Azerbaijan appear to be unstoppable. The sense of despair and defeat that emerged in Armenia from watching these videos was demoralizing. When the battle for the town of Shusha was fought in the first week of November 2020, the conditions for an Azerbaijani victory were set. The result was a decisive military victory by Azerbaijan over Armenia. Decisive victories are rare. We should learn from this war.

Israel utilized a similar information-warfare strategy in the 2021 conflict, as are both sides in the Russia–Ukraine War, but with less success. Some may say information warfare is nothing new, merely the use of propaganda and misinformation to gain an advantage, but now the means have multiplied. The use of streaming video that is self-evident is a powerful means of influence. The Azerbaijanis planned, prepared, and executed their info-war effort in advance of hostilities and then followed through during the fighting as the war produced more content. As a result, they dominated the information-war narrative and began the war by "stripping away enemy leadership options for employing their forces, effectively dominating their decision-making process, not just destroying their assets."[3] This is exactly what the US Army calls decision dominance. As the tempo of war accelerates, leaders must develop and maintain situational awareness faster than ever before. "Speed, range, and convergence give us the decision dominance, and decision dominance gives us the overmatch we need," Army Chief of Staff Gen. James McConville said in a 2021 interview in *Breaking Defense*.[4]

Information war, however, is only a subset of decision dominance. During the initial days of the Israel–Hamas War, the Israelis were able to use information operations to show how they were victims of a massive Hamas rocket barrage aimed at Israeli citizens. This helped to dampen the Hamas narrative that Israel was the aggressor nation. At the same time, the Israelis were able to use information warfare to send messages to Hamas to lure unsuspecting terrorist leaders and forces into coordinated kill zones deep inside their "Metro" system of tunnels. Hamas reacted to Israeli moves and moved on cue while the Israeli leaders and a sophisticated artificial-intelligence coordinated the intricate allocation of sensors and precision fires. This artificial intelligence (AI) aided Israeli decision makers to always be several steps ahead of Hamas as the war played out in the battlespace. The Israelis demonstrated decision dominance over Hamas to such a point that Hamas seemed to dance to the tune of the Israeli AI synchronizing the battle.

In the first weeks of the Russian invasion of Ukraine, the Russians attempted to generate decision dominance through the shock and awe of their first strike. Through this battleshock, the Russians expected to stun the Ukrainians and seize control before they could mount a determined resistance. Russia's plan required a devastating first strike, followed by a rapid advance to secure the levers of power before a stunned Ukrainian population could react. That first strike did not reach the level of disruption the Russians hoped to achieve, and they have not yet gained decision dominance in the war in Ukraine. This is not just a function of technology, but also of the Russians' operational, organizational, and information approach to war.

Singled out was the disruption of Ukrainian internet infrastructure. Russians have always believed controlling the newspapers and radio stations is vital to taking over a country. This was the first thing the Bolsheviks did during the 1917 revolution. Now the infosphere has proliferated to the internet, text services such as Telegram, Reddit, Twitter and Tik-Tok, and instant messaging. Taking control of the internet was an important step to taking over Ukraine. Accordingly, the Russians attacked Ukraine's internet infrastructure, which sustained severe damage during the invasion, but the Russians failed to take control of it, largely thanks to the heroic work of Ukrainian internet providers and Elon Musk's intervention with Starlink. Ukraine rallied, and Ukrainians spread the word of successful defiance to the Russian invasion using their cell phones.

These examples show how the methods of war are changing and why we must adapt to turn these changes to our favor. There are five objectives to win decision dominance: accelerate our decision making (a faster OODA loop) to be able to visualize and act in all domains in real-time; disrupt the enemy's decision-making and communications networks; attack the enemy's plan; shape the battlespace; and seize and maintain the initiative.

Our world is digitally connected as never before, and technologies often transcend their original design. Novel technologies will be used by our opponents to disrupt our networks, as digital communications networks are the new "high ground" of war. Commanders in the battlespace can gain a position of relative advantage with their communication networks, if these networks can work secure and without disruption. Systems such as Joint All Domain Command and Control are evolving to enable a wireless-mesh communications network rather than the point-to-point communications used in the 1990s. A wireless mesh network is a type of network topology where individual devices or nodes connect to one another to form a web-like structure. Data is forwarded between nodes in this kind of network until it reaches its final destination. A wireless mesh network's nodes are usually wirelessly connected, resulting in a highly resilient and scalable network that can be used for a variety of purposes, including data transfer, communication, and internet access in outlying or rural locations. A mesh network is an effective and dependable networking solution since each node can communicate directly with other nodes, the internet, or other external networks, such as satellite communications. A mesh network solves gaps in communications caused by terrain signal blocking. A US Marine Corps program, "Networking on the Move" (NOTM), was used in Afghanistan to provide mobile mesh network coverage in the battlespace. NOTM provided high-bandwidth communications and internet access to marines operating in the battlespace. The system consisted of a series of WiFi-enabled radios mounted on vehicles, which could be linked together to form a self-organizing mesh network.[5]

Decision dominance requires a shared common understanding that graphically represents aggregated battlespace data at all relevant echelons of command and is delivered across a mesh network using the primary, alternate, contingency, and emergency options. AI can be programmed to choose the most effective path to transmit and receive data inside the mesh network on its own. The essence of decision dominance is the ability to perceive and comprehend critical information faster and more clearly than your adversary, and then be able to act on it before the adversary even realizes what is occurring.

In the next few years, AI has the potential to solve many problems to help commanders visualize the battlespace in all domains to secure decision dominance. Decision dominance is as much a commander's ability to gain and maintain the operational initiative by achieving:

> ... a desired state in which a force generates decisions, counters threat information warfare capabilities, strengthens friendly morale and will, and affects threat decision making more effectively than the opponent. Decision dominance requires developing a variety of information advantages relative to that of the threat and then exploiting those advantages to achieve objectives ... The goal is to understand, decide, and act faster and more effectively than the threat. It is not absolute speed that matters, but speed relative to the threat.[6]

Decision dominance is the capability to make better decisions, faster, enhanced by technology and convergence, which is what will set the US military apart from its adversaries. Its importance is magnified by the changing methods of war. Today, with five domains to influence the battlespace, a skilled warfighter can unleash decision dominance in new ways. In addition to the five domains of war (land, sea, air, space, and cyber), there is information warfare and the manipulation of the infosphere. To accomplish this will require a shared understanding across the force, enabled by technologies that empower commanders to see the battlespace, rapidly consider courses of action, make decisions, and issue orders faster than the enemy does. The challenge in today's battlespace, as dramatically demonstrated by recent wars, is that some systems now operate at machine speeds. Galileo once said that "the book of nature is written in mathematics." If so, the book of modern war is written in algorithms, and these algorithms can be harnessed to aid warfighters to see the battlespace and execute Distributed Mission Command in real-time.

Skilled Commanders and Warfighters are Essential to Decision Dominance

As mentioned repeatedly in this book, war is a matter of speed—speed in thinking, deciding, acting, coordinating, and directing—and speed wins war. Speed is the most powerful weapon of war. To achieve decision dominance at machine speeds, the US military must think differently. War will require skilled commanders with many strong operational experiences who can quickly grasp the patterns and counter-patterns of war and use the latest technology to visualize the battlespace in all pertinent domains. Military leaders must be enabled by technology to see and influence the fight in the land, sea, air, cyber, and space domains. Commanders must be able to turn the speed of decision making into a powerful weapon. We must do everything we can to train our leaders and commanders to be excellent at their craft. Training military leaders for command is a key ingredient to gaining decision dominance and this cannot be left to chance or expected to occur during everyday operations.

Great commanders of the past learned the art of war through an iterative experience of combat. Our challenge today is that, to be ready on Day 1, we need commanders who are battle tested at commanding their units without multiple iterations of actual combat. We must use simulation and wargaming to produce commanders with vast tactical, operational, and strategic experience, just as professional sports players are trained and readied for contests. A once-in-a-year battalion, brigade, division or corps war game will not suffice to train battalion, brigade, division and corps commanders for the next war. We need regular training at frequent levels to challenge and attest the warfighting skill of commanders across the force. A significant criterion for selection to command must be a measured skill at warfighting.

All training, short of actual combat, is simulation. Roman legionnaires trained with wooden swords in tough, yet simulated combat. Paper combat-decision games and tactical-decision games, a concept developed in the previous century, are still used today to train and test leaders in tactics and decision making. Today, realistic

Virtual Battlespace Software (VBS) is a tactical military simulation developed by Bohemia Interactive Simulations. VBS is used by the US Army and Marine Corps to conduct constructive simulation at the "crawl" and "walk" levels of proficiency. The unmanned ground vehicle (UGV) shown above depicts a robotic vehicle used in the tactical simulation. VBS4, the latest version of the software, allows users to employ UGVs and test their combat skill on any type of terrain, anywhere in the world, within a high-fidelity, whole-world database. VBS could use existing data from simulations and training events at the Combat Training Centers to create decision aids that would assist commanders in decision making in combat. See https://bisimulations.com/ and https://vbs4.com/. (US Army photo by Sgt Thiem Huynh)

computer simulations dominate the military training arena. All of this is for one purpose—for individuals, leaders, and units to learn before the ultimate test of combat. Live training is the best teacher, but it is also the most expensive. There are never enough resources to do as much live training as you would like. The ideal training program, therefore, maximizes game simulation, constructive and virtual simulations, before advancing to live training. "If you know your enemy and know yourself," Sun Tzu, once said, "you will not be imperiled in a hundred battles … if you do not know your enemy nor yourself, you will be imperiled in every single battle."[7] The idea of a hundred battles is key to Sun Tzu's warning. He was hinting at the number of training exercises that units and leaders must experience to truly know yourself and your enemy. What if you could fight one hundred bloodless battles and learn how to win in a complex world before you ever entered real combat?

The common training paradigm involves four categories of training: live (full-up maneuver, very costly); virtual (using simulated non-lethal weapons or simulated systems, expensive); constructive (primarily focused on teaching the mind to execute procedures and drills, medium cost); and, most recently, gaming simulation (concerned primarily with leader decision making, low cost). In addition, these methods can leverage the latest advances in technology, specifically virtual-reality (VR) and augmented-reality (AR) technologies. VR and AR can increase the realism, immersion, interactivity, and memorability of the training experience. VR is an artificial environment generated by computer hardware and software. It provides the ability to view sensory experiences and synthetic realities and objects on a display, like a computer screen. It encloses the user in a virtual experience. AR is seen through a device to place VR objects in the real world, somehow enhancing the view, as in an aircraft heads-up display. It allows the user to see the real world with added information. AR builds upon the real, physical world by displaying information overlays and digital content tied to physical objects and locations. Let's look at four innovations in simulation that promise to turn innovations into capabilities.

Live Synthetic

There are many military simulations available today, but creating the means to seamlessly combine existing multiple simulations into one holistic training environment has been elusive. This is about to change with the technological fusion of live, virtual, constructive, and gaming simulations into one integrated training environment. The US Army is developing the Future Holistic Training Environment–Live Synthetic, or "Live Synth" to solve this integration challenge. Live Synth is a cloud-based, next-generation approach that will integrate multiple live, virtual, constructive, and game simulations into a holistic, distributed training system. The US Army's National Simulation Center and the US Army Training and Doctrine Command's Capability Manager for Gaming have the mission to make Live Synth a reality between 2023 and 2031. Both of these entities are located at the Combined Arms Center at Fort Leavenworth, Kansas.

When Live Synth goes online, soldiers, aircraft, and military units at installations located anywhere on the globe will be able to connect into a battle network and conduct realistic, immersive, interactive, and memorable training across a wide range of military events and activities. This integrated capability will be brought to the soldiers through a distributed network, rather than bringing soldiers and units from afar to one training area. Live Synth is expected to be a tool that can simultaneously grow individual leaders and train highly effective teams, and do so much cheaper than separate, non-integrated simulations. Combat teams can learn effective use of weaponry, terrain, and increased cooperation in multiple iterations that will increase the level of competency. Leaders, commanders, and staff will be able to plan, prepare, and execute missions. Live Synth will tie all simulations together, allowing for a wide range of full-spectrum combined arms and joint realistic battlespace scenarios.

Constructive Tactical Simulations—Virtual Battlespace

"Virtual Battlespace 3" (VBS3) is a desktop computer tactical training and mission rehearsal, three-dimensional, first-person military training simulation program. VBS3 is developed by Bohemia Interactive Simulations and is the updated version of its successful predecessor, VBS2, used by Australian, Canadian, Finnish, NATO, UK, US, and other militaries for nearly a decade. VBS3 immerses the participants in a fully synthetic and ultra-realistic world that includes weather features, physics-based destructible cover, and realistic weapons and vehicle simulations. This latest system has been optimized to allow for more participants in bigger, more complicated warfighting scenarios. Unlike other simulations, VBS3 allows for many of the nuances of real combat, like suppressive fire and reconnaissance-by-fire, to bring unparalleled realism to a desktop trainer. The simulation can be used to teach small unit-level tactics, techniques, procedures, and doctrine and can be used to test decision making and leadership. It enables soldiers and leaders to conduct a crawl, walk, run training strategy and execute a high frequency of training for low cost. The US Army reports the latest version of VBS3 has increased capabilities that reflect the transition of the Army toward decisive-action operations and an expeditionary mindset. The Army is also adopting the VBS3 game engine as the common engine for the Dismounted Soldier Training System and the Close Combat Tactical Trainer. VBS3 supports more than 102 combined-arms training tasks including:

- Integrate indirect fire support
- Conduct an attack
- Conduct a defense
- Establish an observation post
- Enter and clear a building
- Breach an obstacle
- Conduct route reconnaissance
- Conduct convoy security
- Conduct a roadblock and checkpoint
- Conduct an artillery raid
- Perform tactical air-movement missions

VBS4, the latest software update, improves the previous version and offers a full virtual-desktop training environment with all-terrain rendering for tactical training, experimentation, and mission testing. VBS4 combat scenarios can be fought human against human or human against AI.[8]

Arma 3

"Arma 3" is a Steam-exclusive (digital download) commercial military first-person action game and a very serious military simulation. This game is a commercial spin-off of VBS3 developed by Bohemia Interactive. The large battlegrounds, logistical considerations, and one-hit kills make this Commercial Off the Shelf (COTS) game unique from the normal first-person-action entertainment games. The game is set in the future, during a conflict in the mid-2030s where NATO forces are fighting in Europe against "Eastern" armies led by Iran. In addition, trainers can deploy teams of soldiers to compete against human opponents in force-on-force operations in open gameplay driven scenarios and in massive competitive and co-operative battles online. Arma 3 sells for only US$39 and can be a cost-saving alternative for militaries that cannot afford the more expensive VBS3 system. To enhance the training value of Arma 3, trainers could employ human observer-controllers to participate as role players to augment the simulation and amplify the training objectives. Training with Arma 3 could also be enhanced by VR technology.

The Rift and the Quest

Oculus (formerly Oculus VR) is a Menlo Park, California, technology company that started as a crowdfunded Kickstarter startup in 2011 and now belongs to Meta (Facebook). Oculus Rift, first released in 2016, is a VR headset developed and manufactured by Oculus. It is designed to provide an immersive experience for gamers and other users by providing a 360-degree VR experience. A new version, called the Meta Quest 2, is now available and is already being adopted for military simulations. Oculus has described it as "the first really professional PC-based VR headset." This headset is a VR system that will also work with many Android mobile devices. Combining the Quest 2 with the computer power of mobile devices provides a powerful combination. The current generation learns on their mobile devices. Exploiting the power of distributed and mobile devices to access training content is a critical part of any future acquisition and training strategy. This high-definition (HD) headset can turn a laptop, tablet or mobile phone into a portable next-generation VR system. The Rift steps the player into the VR world, immersing them in a wide field of view. The device creates a sense of presence and immersion by using lenses and sensors to track head movements, allowing users to see the virtual environment from different angles. The combination of the wide field of view and stereo-earphones makes you feel as if you are actually in the synthetic VR world. The Quest has a very high refresh rate and a low-persistence display that uses state-of-the-art optics designed specifically for VR and is currently supported by Unity and Unreal game engines.

The older Rift is used by several military organizations and even the US Secret Service. The US, UK, Dutch, and other military forces use the Rift for military simulations. The Norwegian Army has even used the Rift as their viewer for a see-through armor set up for armored vehicles.[9] The US Navy is using the Rift to train sailors in future warfare in a program designed to enhance communication

Military units do not fight to the level of their leader's expectations, they fall to the level of their collective and individual training. Simulations are a way to increase mastery of individual and collective training skills. Virtual Battlespace 3 is a constructive simulation of a realistic semi-immersive environment allowing military units, mainly at company level and below, to practice more than 150 combat exercises, platoon-level collective tasks, combined-maneuver tasks, and other collective tasks. (US Army image)

and collaboration. This US Navy program is *Blueshark*, a joint initiative between the Swampworks division of the Office of Naval Research and the University of Southern California's Institute for Creative Technologies. The Norwegian Army is testing the Rift's application in tank-driving scenarios. In January 2015, the British Army used a prototype Rift to allow civilians to experience driving armored vehicles in an effort to entice new recruits to sign up for service. The Rift is also being considered by the US Air Force for use by unmanned aircraft system pilots to make flying easier and more accurate.

The typical 3D HD VR headset costs about US$500. As the Quest and other head-mounted displays improve, become smaller, and less expensive, VR technology could quickly revolutionize the world of military training simulation. Combining commercial VR systems like the Quest and government simulation solutions, like VBS4, can create a distributed training environment that will help to develop leaders who can lead more effectively, fight more efficiently, and learn to adapt, improvise, and overcome.

In this photo, US soldiers train with the Integrated Visual Augmentation System (IVAS) as a part of Project *Convergence 2022* at Camp Talega, California, on October 11, 2022. At the core of IVAS is the Microsoft HoloLens. The HoloLens is a powerful mixed-reality headset that allows soldiers and commanders to collaborate "in the moment" from anywhere, as long as they are connected to the network. The Microsoft HoloLens and IVAS have possibilities for unit command and control that should be explored. (US Army photo by Sgt Thiem Huynh)

If you want your commanders and leaders to be "off the ramp and ready" capable and trained to generate overmatch in any battle situation, you must "train to habit." Training to habit transforms knowledge into capability and requires multiple iterations of immersive, interactive, and memorable training experiences. The latest military simulations can provide unparalleled realism and training benefit and generate cost-effective training with minimum risk. The challenge today, in a time of shrinking military training budgets, is to develop a training-simulation acquisition strategy that provides the most benefit within funding limitations. This is a tall ask indeed.

When budgets get tight, trainers must focus on the most vital training needs. Leader development is the most cost-effective investment to improve a military force. Leaders can turn innovation into capabilities. Good leaders create effective units, breach capability gaps with tactics, techniques and procedures, and find ways to overcome obstacles and win. Simulations, especially constructive and game simulations, can be very effective tools in developing leaders. Our goal should be to have leaders fight a hundred bloodless battles in live, virtual, constructive, and game simulation, before the first combat. Even in an era of technological warfare, the human element is predominant.

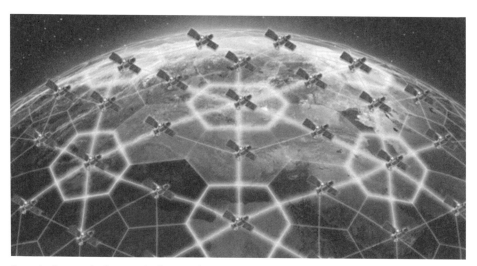

Space is the new high ground of war. Who controls the stratosphere and space, controls the earth. The US Defense Advanced Research Projects Agency (DARPA) is developing a program called Space Based Adaptive Communications Node (Space-BACN) to create a low-cost, reconfigurable optical communications terminal that adapts to most optical intersatellite link standards, translating between diverse satellite constellations. Space-BACN will connect existing satellites in new ways and will support US military programs such as the Joint All-Domain Command and Control (JADC2) system. SpaceX is one of the teams working with DARPA to make this a reality. (DARPA image)

CHAPTER ELEVEN

The First Starlink War

> Orbital position and frequency are rare strategic resources in space. At present, the geosynchronous orbit has almost been fully occupied and the scramble for Low-earth Orbit and Medium-earth Orbit positions has become more intense. The LEO is able to accommodate about 50,000 satellites, over 80% of which would be taken by Starlink if the program were to launch 42,000 satellites as it has planned. SpaceX is undertaking an enclosure movement in space to take a vantage position and monopolize strategic resources.[1]
>
> CHINA MILITARY ONLINE, MAY 12, 2022

Lieutenant Colonel Wang Liang stands on a helicopter landing pad in the People's Liberation Army camp near the town of Mumian on the island of Hainan. The pad is a new addition to the facility. It has one purpose: to launch high-flying aerostats.

He watches in anticipation as six soldiers comprising the ground-launch crew fill the balloon with helium. Eventually, the balloon is full, extending nearly 200 feet tall.

Liang thinks the sight is magnificent. The upside-down teardrop shape, white against a deep-blue sky, seems like a beautiful painting. For a moment, he forgets this is a high-tech weapon of war.

The ground crew connects a long line that is laid out on the helicopter pad. The sturdy line, made of a flexible metallic cable, holds the balloon to the ground.

The ground crew lifts the cargo package up as the lines are slowly let loose. As the balloon gains altitude, another crew of 20 soldiers lifts the solar panels off the helipad, 10 men to each side, careful to make sure the ropes do not snag as the balloon pulls the sensitive cargo package into the air. As the balloon drifts skyward, the rope connecting the solar-panel array goes taught. The ground crew releases the cargo package and in seconds the balloon, its sensitive cargo, the solar panels, and the electric turbofan motors that will steer the balloon, are soon a dozen feet above the pad.

Liang uses an Android tablet controller to start the balloon's propellers. The motors start and a dozen small turbofans buzz. Located at the bottom of the solar-panel array, these fans will guide the weapon. Soon the balloon is taken by the wind, floats northeast, towards its rendezvous with the enemy. He watches as the balloon rises higher and then looks back at General Qiao Xiangsui.

"This phase is done," says the general. "Let's move to the monitoring station."

Liang complies. He places the tablet in his left hand and walks by the general's side, one step behind.

"You know, the Germans did the exact same thing right before World War II," the general declares. "In 1939, they sent the Graf Zeppelin, their most sophisticated lighter-than-air craft, filled with the latest electronic-intelligence gathering devices, to conduct an electronic reconnaissance of the British Isles and test the British radar."

The younger officer nods in agreement and continues walking at the general's side.

"What do you think, Liang, of our latest aerostat?" the general asks as he strolls on.

"It is another wise decision of the great Chinese Communist Party," Jiang replies, providing the answer he thinks the general is searching for.

The general abruptly stops and turns. Liang halts and stands at attention.

"You, my young colonel, are expected to think. I am not training you to be a parrot."

Liang takes a deep breath and nods.

"I ask you again and I want an intelligent answer," the general orders. "What do you think of our weapon?"

"This aerostat will complement our other means to gain dominance of the enemy's adjacent space, that area of space that is between 15 kilometers and 50 kilometers above the range of aircraft and below that of low-earth-orbiting satellites."

"Yes, any fool knows this," the general retorts. "But why use a balloon and not an aircraft or satellite?"

"Our weapon is a very sophisticated near-space vehicle," Liang answers, summoning his courage. "The radar and infrared signatures are not obvious, so it is difficult for enemy sensors to detect. Most enemy combat aircraft and surface-to-air missiles cannot reach near-space and, if it is detected, the aerostat may not look like a threat. We can tell everyone it is merely a civilian weather balloon gone off course. Most importantly, near-space vehicles such as our balloon can operate over a single area for a long time as we gather all the intelligence information we need."

The general nods in agreement. "It seems I have not wasted my time on mentoring you." He looks up as the balloon sails out of sight. "Anything else?"

"Yes," Liang answered. "The emergence of new technologies has presented endless possibilities for match ups involving various old and new means to combine and provide ways for us to overcome our enemies."

"Good. Almost an exact quote from *Unrestricted Warfare*,"[2] the general replies. Then, with a slight grin, he asks: "How long will it take for our aerostat to reach the target?"

"Sir! It will be over Taiwan in two days."

"Imagine if I had a hundred such aerostats, loaded with quantum sensors and powerful weapons," the general muses, sounding almost gleeful. "Our enemies are thinking two-dimensionally. The stratosphere is more decisive terrain than

High-flying stratospheric balloons have an important role to play in the provision of military surveillance, electronic warfare, communications, and as weapons platforms. In this photo taken on February 3, 2023, a US Air Force pilot looks down on a Chinese surveillance balloon that was flying over the US for six days. On February 4, 2023, the US Air Force shot the balloon down off the eastern coast of South Carolina. (US Department of Defense photo)

low-earth orbit. We will use these aerostats, not only to gather intelligence and reconnaissance information, but also as electronic jamming and weapons platforms. Once over the enemy's country, if several were armed with electromagnetic-pulse bombs, we could detonate them simultaneously over the enemy's territory to destroy their communications and electronics, leaving them blind, confused, in the dark, and easy prey."

Liang listens intently. He knows that when the general starts talking like this, it is not his place to interrupt or comment.

"Imagine. Someday, at the cost of polyurethane and helium, and a few exquisite pieces of technology, I will dominate skies over America with such aerostats," the general bragged. "Just wait and see."

***

The story above **might** have been the conversation as the Chinese People's Liberation Army (PLA) launched intelligence, surveillance, and reconnaissance (ISR) aerostats over Taiwan in March 2022.[3] This fictional dialogue is based upon PLA articles that describe how stratospheric balloons can be used as communications nodes, sensor platforms, and, if properly equipped, as weapons. We will never know, as the Chinese are not about to broadcast their intentions, but their actions are clear. On January 28, 2023, a Chinese ISR aerostat entered US airspace, to the amazement of the American people and the befuddlement of the Department of Defense.[4] On February 4, the balloon was shot down by a missile launched from an F-22. The Chinese government protested the destruction of their aerostat and accused the US of "overreacting" and "seriously violating international practice."[5] The US government blacklisted several Chinese aviation companies and sent strongly worded messages to the Chinese but took no further action. With such a vigorous reaction to the Chinese violating US airspace, it would not be surprising to see more high-tech aerostats used for military and intelligence-gathering operations in the future.

The Chinese do not have the ability to launch as many ISR satellites as they would wish, nor thousands of small satellites into low-earth orbit as Elon Musk has done since 2018, but they are applying different means to accomplish the same goal: deploy systems to dominate the stratosphere, the edge of space. On the other side of the globe, in the ongoing Russia–Ukraine War, the need to reestablish communications

A Falcon 9 rocket carrying the Starlink 4-37 payload lifts off from Space Launch Complex 39A at Kennedy Space Center, Florida, on December 17, 2022. Starlink is the satellite network developed by SpaceX to provide low-cost access to the internet and remote location connectivity. Starlink plays a key role in the war in Ukraine, connecting Ukrainian fighters to the internet. (US Space Force photo by Joshua Conti)

networks once the Russians smashed and made the Ukrainian internet ineffective, became a priority. Move, shoot, and communicate—these are the primary functions for all military units in combat. All three are critical, but in our technology driven age, communication is transcendent. If you cannot communicate by radio or digital means, you cannot effectively coordinate your striking power nor execute combined-arms operations or cross-domain maneuver. Take away the enemy's ability to communicate and you negate his ability to move and shoot effectively.

An ongoing observation from the Russia–Ukraine War is the importance to communicate while under electronic-warfare (EW) and cyberwar attack. As a precursor to their ground invasion on February 24, on February 15, 2022, the Russians executed extensive cyberattacks on Ukraine's internet. These included continual Distributed Denial of Service (DDoS) attacks culminating on the day of the invasion, when Ukraine's primary internet-service provider, Triolan, was taken down by Russian DDoS offensive actions. Simultaneously, the Russians executed targeted kinetic, EW, and cyber strikes on Ukraine's internet infrastructure. These DDoS attacks caused internet outages across large swathes of the country and left thousands of Ukrainians without communications and in the dark about the Russian

US airmen set up Starlink terminals at Kadena Air Base, Japan, January 10, 2023. The Starlink terminals, performing at maximum speeds, are 4,000 percent faster than previously used antennas, which allows for ease of communication for connected airmen. Similar terminals were shipped to Ukraine by Starlink to support Ukraine's war effort and reestablish internet service. (US Air Force photo by Airman 1st Class Sebastian Romawac)

invasion, creating fear among the citizens. This lack of communications not only denigrated the ability of the Ukrainians to mount an organized resistance, it also fanned the flames of panic. Despite their initial success, the Russians did not achieve their strategic goal of eliminating Ukraine's internet. One reason for this resilience: Elon Musk's Starlink.

What is Starlink?

Starlink is a next-generation satellite network owned by SpaceX. It provides broadband internet access to enable voice and data communications anywhere in the world. A constellation of thousands of satellites in low-earth orbit, circling the earth at a distance of about five hundred fifty kilometers (three hundred forty-two miles), beams the signal from space to terrestrial Starlink receivers. Most single geostationary satellites orbit Earth at about thirty-five thousand kilometers (twenty-one thousand, seven hundred fifty miles). Since Starlink satellites are much closer to ground-based stations, the data transmission time is faster, and this generates high-speed data transmission for earthbound devices. All that is needed to access Starlink is a receiver and a subscription. It also works on the move, an important consideration for military use.

SpaceX launched the first micro-satellites in February 2018. Starlink has launched thousands of satellites into space to make up a constellation of internet routers orbiting the globe in low-earth orbit. These satellites are small when compared to the size of most telecommunication satellites—approximately 3.2 m × 1.6 m × 0.2 m in size and weigh about two hundred twenty-seven kilograms (five hundred pounds). Each satellite has a solar-panel array to generate power to run the on-board computer and electronics. Four phased-array antennas and two parabolic antennas transmit and receive data. A unique ion-propulsion system fueled by krypton, the first such ion–krypton drive to operate in space, powers each satellite. Starlink satellites have a built-in, autonomous collision-avoidance system that automatically maneuvers the satellite to avoid hitting other objects in space. The on-board navigation system scans the stars to ensure orientation, altitude, and location to generate the optimum signal to ground receivers. To improve communications security and speed up data transmission, an optical space laser system was added to every Starlink satellite launched after January 24, 2021. Starlink currently provides broadband internet service to 32 countries around the world.

The Ukrainians Use Starlink for War

On February 26, 2022, two days after the start of the Russian invasion, in an open plea to Musk on Twitter, a Ukrainian government official asked him to restore critical internet services for their military and government agencies by lending them his

Starlink network. "While you try to colonize Mars—Russia try to occupy Ukraine! While your rockets successfully land from space—Russian rockets attack Ukrainian civil people! We ask you to provide Ukraine with Starlink stations and to address sane Russians to stand," tweeted Mykhailo Fedorov, Ukraine's vice-prime minister and minister of digital transformation.[6]

Musk immediately sent Starlink receivers to Ukraine. "Starlink—here," Fedorov tweeted on February 28, along with a photo of the Starlink receivers. "Thanks, @elonmusk." Musk responded, "You are most welcome." By the first week of March, Musk had restored internet communications for Ukraine. Starlink now connected units of the Ukrainian Army. The Russians immediately scrambled to attack this alternative source of communication. "Talked to @elonmusk. I'm grateful to him for supporting Ukraine with words and deeds. Next week we will receive another batch of Starlink systems for destroyed cities,"[7] tweeted President Volodymyr Zelensky on March 5, 2022. Musk foresaw this Russian reaction to disrupt the Starlink network. "Starlink is the only non-Russian communications system still working in some parts of Ukraine, so probability of being targeted is high. Please use with caution,"[8] he tweeted in reply. Russian hackers attempted to disrupt the system, but with limited success. On March 26, 2022, Musk tweeted again: "Starlink, at least so far, has resisted all hacking and jamming attempts."[9] As is usual in war, a race was on between Musk's software engineers defending their system and Russian cyber-attackers attempting to bring it down.

The Ukrainians learned they could use the Starlink network for more than just voice and data transmissions. Starlink can also communicate with drones. In the early days of the war, the Ukrainians successfully employed drones, particularly unmanned combat aerial vehicles (UCAV) and loitering munitions, to find and strike Russian command posts, artillery, and high-value targets. As the fighting wore on, the Russians adapted, massing their EW systems and focusing on cyberwar efforts to jam Ukrainian drone signals. Jamming line-of-sight signals from terrestrial controllers is the mission of many counter-unmanned aerial system weapons and has proven to be an effective method in disrupting drone operations, particularly small unmanned aerial systems. Ukrainian operators quickly learned they needed to improvise to overcome Russian jamming. Aerorozvidka, the Ukrainian Army's drone unit founded by volunteer internet and drone experts, adapted Starlink to control their drones. "We use Starlink equipment and connect the drone team with our artillery team," said Yaroslav Honchar, an Aerorozvidka commander in an interview in the March 18, 2022, edition of *The Times*. "If we use a drone with thermal vision at night, the drone must connect through Starlink to the artillery guy and create target acquisition."[10]

With Starlink, Ukrainians regained access to the internet and successfully countered Russian EW jamming. Gwynne Shotwell, the president and chief operating officer of SpaceX, said providing Ukraine with Starlink broadband internet services

Ukrainians regained access to the internet and countered Russian electronic-warfare (EW) jamming once Starlink was deployed in Ukraine. Starlink is used by Ukraine's elite drone unit to communicate with small unmanned aerial systems to target Russian soldiers and equipment. SpaceX has donated 3,667 terminals worth around ten million dollars. (Image posted on Twitter on March 9, 2022, by Mykhailo Fedorov Minister of Digital Transformation of Ukraine)

was "the right thing to do ... the best way to uphold democracies is to make sure we all understand what the truth is." On July 25, 2022, Fedorov tweeted Musk again: "As of now, we have over 12K [12,000] Starlinks in Ukraine. They help with access to high-quality internet in difficult conditions. Still, we are wondering of all possible ways to use @SpaceX tech. Our telecom team made Starlink test drive on the road at a speed of 130 km/h."[11]

Starlink is not a "silver-bullet" solution, but it has proven to be extremely resilient in combat despite the Russians' best efforts to use their cyber and EW capabilities to defeat it. As the fighting continues in Ukraine, other nations are taking notice of Starlink's effectiveness as a wartime means of communication. According to *Business Insider*, in an article by Ren Yuanzhen in *Modern Defense Technology* magazine, published in China, the PLA is studying ways to counter Starlink. The PLA understands Starlink's potential and considers it a major threat. "It is recommended to apply a combination of soft and hard killing methods to disable some Starlink satellites and destroy their operating systems." The Russians concur with the PLA's assessment. Dmitry Litovkin, editor-in-chief of the *Independent Military Review*, as quoted in *Radio Sputnik* on May 9, 2022, stated, "The Starlink constellation of 2,000 satellites has performed well in Ukraine. China sees it as a threat to its security. It is capable of transmitting data from aircraft and drones a hundred times faster. Therefore, in Beijing, they want to create a system that would be able to track and destroy satellites."[12] In an article published in *China Military Online* in May 2022, Li Xiaoli posited:

> When completed, Starlink satellites can be mounted with reconnaissance, navigation and meteorological devices to further enhance the US military's combat capability in such areas as reconnaissance, remote sensing, communications relay, navigation and positioning, attack and collision, and space sheltering. Clearly, the military applications of the Starlink program will give the US military a head-start on the future battlefield and become an "accomplice" for the US to continue to dominate the space.[13]

The Chinese and Russians have much to be concerned about. While China is working to put satellites in orbit, neither they, nor the Russians, have their own Starlink constellation. Elon Musk's entrepreneur-to-government cooperation frightens them. On March 31, 2022, the United States Air Force embedded a Starlink gateway router into the travel pod of an F-35A Lightning II. An Air Force crew linked the signal to a Starlink receiver on the ground. The test proved Starlink could send data 30 times faster than traditional connections. Starlink can also provide secure communications links to unmanned aerial systems, potentially enabling communications for drone swarms. As unmanned systems warfare increases in scale, tactical employment, and capabilities, satellite communications systems like Starlink will be an indispensable element in modern warfare. Brandon Wall and Nicholas Ayrton, in an article in the US *Center for International Maritime Security* published in September 2021, hypothesized, "A small fleet of semi-autonomous drones could also act as a screening force for operations, acting to provide an extended sensor net and provide greater tactical awareness, be they for combat operations or as an early warning system."[14] Starlink could provide the means to control these drones, anywhere on the planet. With a successful launch on March 24, 2023, SpaceX had 4,161 Starlink satellites in orbit with more to be launched in the months to follow.[15] There is also a drone version of Starlink as a Canadian company called RDARS successfully integrated Starlink into its drone system in November 2022. A drone version would provide on-call coverage in remote areas where Starlink is unavailable or under attack by EW.[16]

SpaceX presented Starshield, a Starlink for the military, in December 2022. This is particularly important as the Russians are now using sophisticated signals-intelligence equipment and EW systems to jam or locate Starlink's broadband internet signal in the battlespace in Ukraine. Elon Musk's team is coming up with fresh solutions to deal with this by developing Starshield. It will provide a secure satellite network for the US government and military. "Starshield leverages SpaceX's Starlink technology and launch capability to support national security efforts. While Starlink is designed for consumer and commercial use, Starshield is designed for government use, with an initial focus on three areas: earth observation, communications, and host payloads."[17] Starshield will include inter-satellite laser communications links to connect non-Starlink satellites into the Starlink network.

The ability to put a constellation of small internet satellites in low-earth orbit, on demand, is a unique capability that can enable terrestrial manned and unmanned combat operations. With the Ukrainians using Starlink to combat the Russians, essentially fighting the "First Starlink War," Moscow is feeling the pain. SpaceX and Starlink are a fascinating example of how industry, with the right leadership, is deploying technology that is changing the methods of war.

CHAPTER TWELVE

Preparing for the Next City Fight

The complications Russia has encountered in urban conflict in Ukraine's Mariupol, Kharkiv, and Kyiv are not simply a function of Russian incompetence. They're a reflection of the difficulties any military would face in urban warfare.[1]

MARGARITA KONAEV AND KIRSTIN J. H. BRATHWAITE, IN A REPORT ON RUSSIA'S DIFFICULTY IN WINNING THE URBAN BATTLE

"No grenades for the drone," a young Ukrainian soldier offered. He is wearing a white snow-camouflage uniform, a helmet, and carrying a rifle. Blue tape wrapped around his arms identifies him as a Ukrainian soldier. On the left shoulder of his uniform is the patch of the Ukrainian Aerorozvidka (Aerial Reconnaissance) unit. Only 11 months ago, he was working as an information technology department administrator in an office in Kharkiv. "But I can place it on the roof to overlook the road."

"Do it," the sergeant, named Mykola, replies as he takes one last puff from a cigarette and then puts it out on the windowsill. Mykola scans the street. Five hundred meters in front of him is a burned-out Skoda sedan lying upside down in the center of the road. His four-man detachment, consisting of two drone operators, one sniper, and himself, are to slow down the enemy's advance. "The Orcs are on the move. I can feel it."

Four propellers buzz as the small unmanned aerial system, a DJI Phantom-4 quadcopter, lifts off the floor and flies through an opening in the wall of the battered building. Flying low for a few minutes, the drone then pops up to the roof of a burned-out building. A Ukrainian drone operator, a young man of 23 named Yuriy, controls the quadcopter's movement using a tablet. Carefully, he lands the drone on the roof, turns off the rotors, and aims the drone's camera.

"Our drone is set," Yuriy reports as he stares at the tablet.

Machine-gun fire echoes from somewhere in the distance. The city is in ruins. Debris lines both sides of the street. Several three-story apartments in an apartment complex on the right side of the road are smoldering from fires started by recent artillery strikes.

"Here they come," Yuriy announces. "There are four. The last one has a grenade launcher."

A Ukrainian soldier stands guard near Bakhmut on December 3, 2022. Bakhmut was one of the fiercest and bloodiest battles of the Russia–Ukraine War. "There was no place left in Bakhmut that was not covered in blood," President of Ukraine Volodymyr Zelensky wrote in his Telegram channel. "There is not an hour when the terrible roar of artillery does not sound." (mil.gov.ua)

"Four?" Mykola questions. He knows he is lucky to have the drone team with him today. The drone can attack with grenades or be a robotic scout, providing early warning. There are too few drones, and they seldom survive two or three missions. Their survivability rate, he muses, is not much better than that of a soldier in this hellish place. "Wagnerites. They usually come in a group of 10."

"Only four," Yuriy confirms.

"Wagnerites brag they will all go to hell, but in hell they will be the best," Mykola comments, looking at the sniper. "Let's give them what they ask for, a trip to hell."

Another young Ukrainian soldier picks up a Dragunov SVD63 sniper rifle and takes a firing position inside the building. Propping his rifle on a table, he has a perfect view of the street through a broken window. Shooting from inside the room, and not sticking the rifle's barrel outside the opening, should mask the flash of his fire from the approaching enemy.

Mykola grunts.

"They're to the right of the Skoda," the drone operator reports.

"All four of them?" Mykola asks.

"*Tak*, they're about to turn the corner."

"Any sign of tanks or BMPs?" Mykola inquires.

"No, they appear alone," Yuriy answers. "Most likely a recon patrol."

"I see the lead man," the sniper replies.

"Fire when the last one passes the car. Shoot the Orc with the grenade launcher," Mykola orders, then turns to Yuriy. "As soon as the shooting starts, get the drone out of there and move back to our next position."

Yuriy nods.

"Don't miss," Mykola says to the sniper. "We won't get another chance."

"I never miss," the sniper replies with a smile.

Seconds pass. Four Russians move past the Skoda.

The Ukrainian sniper fires.

Time freezes for a moment for Mykola as he sees the Russian carrying the RPG (rocket-propelled grenade) crumple to the ground. Three other Russians hit the pavement and return fire, but their shots fly wild as they cannot identify the sniper's location. These are Wagner mercenaries, who know a bit about war, but it is all happening too fast.

Mykola fires his AK-74 assault rifle at the Russians as the sniper aims at the next target.

The Russians crawl away from the fire, hoping to use the Skoda as cover.

The sniper hits another Russian. Two are now down.

One Russian blasts away with his rifle and then rolls behind the car. The other Russian drags a wounded man behind the Skoda and out of view.

"Cover your ears," Mykola announces, as nonchalantly as if he was asking the sniper to pass the salt to him at the lunch table. He puts down his Kalashnikov, picks up a detonator, then squeezes the clacker twice.

A bright flash as the explosive hidden under the Skoda detonates. Fire, smoke, and chunks of car hurtle skyward. As the scene clears, Mykola sees no sign of the Wagnerites.

Shrieks of incoming shells fill the air. The floor tremors as the adjoining building is hit. Russian artillery. The drone team runs out the back door as shells explode. Inside their room, steel fragments hit the wall and pieces of the roof fall.

The building fills with dust and smoke.

"Time to go," Mykola orders. "Run!"

The Russian artillery is accurate. A Russian artillery observer or a Russian drone must be able to see them. Mykola knows he has seconds to run or die. More shells slam into buildings nearby. They exit out the back door. As they run, Mykola sees the dead body of Yuriy and his assistant, both killed by Russian artillery.

Mykola continues as more shells smash into the building behind him. "Another day in Bakhmut," he thinks as he races across the courtyard. The Russians are pressing hard. How many more days of this can we take?

***

The urban battles in Ukraine have dramatically demonstrated the need to prepare and train ground forces for combat in cities. When battles are inevitable in densely populated cities, new methods of fighting must be devised. Room-clearing drills are familiar to US infantry units, but the US military's operational and strategic thinking regarding urban combat is deficient. (US Army photo)

The story above is fictional but is based on actual videos of the fighting in Bakhmut, Ukraine.[2] Urban warfare is today and tomorrow's war. As the Russia–Ukraine War grinds on, towns, transportation hubs, and population centers are key terrain in the modern battlespace. Fighting in cities such as Kyiv, Kharkov, Mariupol, and Bakhmut occurs because these cities control the roads. Armies must move on roads. In Ukraine, the thick, *Rasputitsa* (Russian) or *Bezdorizhzhia* (Ukrainian)[3] mud of autumn and spring makes traveling cross-country difficult for infantry and nearly impossible for vehicles. Tanks and infantry fighting vehicles traveling off the roads sink into the gooey mud and are immobilized. The roads are vital to move troops, vehicles, equipment, and supplies, and the cities control the roads.

The most significant wars in the past three years were fought in urban areas and are the harbingers of battles to come. Azerbaijan fought Armenian forces in the Second Nagorno-Karabakh War, from September 27 to November 10, 2020, and clinched a military victory when Azerbaijani special forces and light infantry, supported by artillery and drones, conducted a successful infiltration attack to seize the city of Shusha, which is the decisive terrain in Nagorno-Karabakh. The Israel–Hamas War, May 11–21, 2021, is an exemplar of how to wage modern urban combat as the battle was fought in the city of Gaza and the tunnels beneath the city. Gaza has an estimated population of 750,000 and the World Population Review website named it as one of the most densely populated cities on earth. Russia's invasion of Ukraine on February 24, 2022, which kicked-off the ongoing Russia–Ukraine War, is fought over cities, towns, and villages. Battles such as the siege of Mariupol, where outnumbered and outgunned Ukrainian soldiers defended the city from February 24 until May 20, 2022, is just one example of urban combat in modern war. As military leaders and weapons-system designers survey these conflicts, the pressing need to study, train, equip, and prepare for urban warfare is obvious and urgent. In modern war it is impossible to avoid battles in cities and megacities. These complex battlefields absorb combat power like a sponge soaks up blood. Military strategists from Sun Tzu to the current day have warned commanders to avoid urban combat.[4] Today, 57 percent of the world's population lives in urban areas. Megacities, with populations over 10 million, are the focus of political, economic, military, and human power. "By 2030, cities will account for 60 percent of the world's population and 70 percent of the world's GDP [Gross Domestic Product]."[5] In addition to their populations, modern cities provide a complex battlespace that is easy to defend. Most cities have buildings made of steel and reinforced concrete that are 20–70, or more, stories tall. Fighting to take these types of buildings, filled with defenders determined to die for every floor, every building and every block, requires a tremendous investment in combat power and results in heavy casualties on both sides, and for the civilian population. Contrary to Sun Tzu, sieges may become the preferred tactic, when laying siege is an option, as the heavy cost of fighting in modern megacities is prohibitive.

Urban terrain poses extreme challenges to military operations, diluting the range, precision, sensing, communication, and stand-off protection advantage of modern military forces. Belligerents can mask in a city better than in most other terrain. There are no easy, high-tech, bloodless solutions to this kind of battle, if fought in the old way, but new technologies in the hands of well-trained combined-arms units can provide a winning edge. For the foreseeable future, urban combat will remain an infantry centric affair, but mobile striking power, sensors and strikers, and dominating the space above a city, hold the keys to enabling infantry to win the city fight.

Mobile Striking Power

Today, the central element of any combined-arms team is the main battle tank (MBT). You cannot fight effective combined-arms combat in urban areas without them. Many articles were published in past decades about the "death of the tank,"[6] but there is no replacement for the mobile, precision firepower, command-and-control networking, and protection that tanks provide. Current tanks, however, lack four important capabilities to generate mobile striking power in the modern battlespace. Mobile striking power is the offensive capacity of any military system, unit, or force to generate offensive action to move across the battlespace to disable or destroy the enemy.

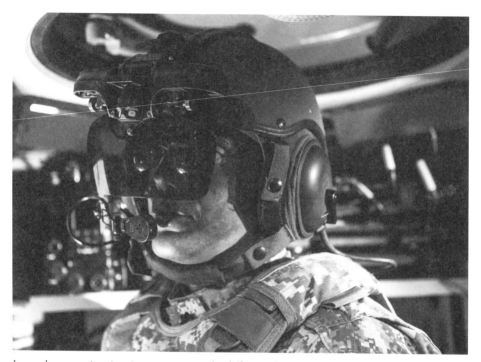

In combat, superior situation awareness is the difference between life and death, especially for tank commanders. A tank commander that can see through the tank's armor would have a tremendous edge in a close-in fight. Elbit System's helmet-mounted IronVision "HMS" combines a high-tech helmet with external cameras placed on the outside of the vehicle to create a virtual view of the battlespace. (Elbit Systems)

For over a hundred years, tanks have performed this role. Legacy MBTs operate best in open terrain to engage targets in long-range combat and are not optimized for the close-in battle of the urban canyon. An urban canyon is a place where the streets are flanked by tall concrete and steel buildings on both sides, creating a man-made canyon-like environment. In this terrain, tanks and armored infantry fighting vehicles are channelized and become easy prey for short-range anti-tank weapons. In the Russia–Ukraine War, the lesson of tanks in urban combat is not that tanks are obsolete, but that they were not employed correctly. Effective tank tactics require combined arms in every type of terrain. The increased effectiveness of long-range precision fires, combined with multidomain sensor networks easily defeat a single-arms approach. In short, legacy tanks without combined arms are a recipe for disaster in the urban battle.

Cold War-era tanks have a maximum main-gun elevation of $-10°$ to $+60°$ (which means they have trouble hitting targets on the fourth floor of a building at close range), are difficult to mask from enemy sensor networks, have limited connectivity to a larger warfighting network, and poor situational awareness (SA)—crewmen are unable to see the full battlespace when they are "buttoned-up" and under armor. SA is key to surviving and winning. Tank commanders must be able to observe, orient, decide, and act more rapidly than their threat. In tank battles, the tank commander who sees, identifies, and shoots first, usually wins. This is difficult in the confusion and chaos of close-in urban combat. When a tank crew is buttoned up and operating under armor, they see through vision blocks and narrow gun sights; their SA is low. To say their vision is restricted is an understatement.

Technology can change this. Four major improvements to tanks can turn these deficits into advantages. The first is enhanced SA. Second is improved survivability by upgrading tanks with new designs and added active and passive defensive systems. Third is to enable tanks to work seamlessly to command ground robotic systems and aerial drones as "loyal wingmen" to reinforce combined arms. Fourth is to embed in each tank the computing and networking power to enable the tank to act as a command-and-control node in a tactical mesh network.

See-through armor[7] systems enable crewmembers to enhance SA by seeing through an array of cameras. With see-through armor, it appears to the tank crew as if their heavy protection is not there at all. Just as a reversing camera on a car provides the driver with enhanced driving ability, see-through armor increases crew decision making and reaction speed. Next-generation armored vehicles updated with see-through armor are maximized for closed-hatch operations.

Enhancing protection of modern tanks and armored fighting vehicles (AFV) will require placing the crew in the hull, where the armor can be the strongest, and replacing manned turrets with robotic turrets. A robotic turret with an automatic

There is no need for tanks in the modern battlespace, unless you want to win. Tanks are essential for offensive operations. New designs, like this screen capture of a General Dynamics Land Systems AbramsX, are examples of what can be done to increase the survivability, lethality, and connectivity of main battle tanks. The AbramsX future tank design is artificial intelligence-enabled, has a hybrid electric engine, a robotic turret with an automatic cannon, and a crew of three located in the hull. (General Dynamics Land Systems)

loader for the main gun can also reduce the weight of the tank or AFV, making it easier to deploy operationally and strategically. Designing the system with a hybrid electric engine will not only improve fuel efficiency and operational range but provide the electricity required to operate electronic computing systems on board the vehicle.

Upgrading existing and new platforms with the ability to command—not control—robotic systems is the next step in armored vehicle evolution. Future AFVs will be designed to command swarms of unmanned aerial systems (UAS) and unmanned ground vehicles (UGV). Producing smaller, lighter, faster, cheaper, unmanned robotic tanks, designed to be commanded by human tank commanders using voice commands, are needed. In such an arrangement, the military robot would execute a set of battle drills to engage the enemy and would do so on human command, executing the drill autonomously until redirected by the human to stop or execute another drill. Voice command of robots is possible, as simplistic systems such as Amazon's "Alexa" show. Today, many people believe all robotic systems must be controlled by a human-on-the-loop, with a human hand on a remote controller and trigger, but will this debate end when robots with OODA (observe, orient, decide, and act) loops of nanoseconds start killing humans with OODA loops of milliseconds?

Unmanned light tanks, like the M5 Ripsaw[8] designed and built by Howe & Howe Technologies/Textron, demonstrate a robotic system to support and reinforce human-operated MBTs that could be adapted to human command. Weighing only 9,000 pounds (4,100 kilograms) and able to reach speeds of 65 mph (105 km/h), the M5 is an all-electric tank, armed with a 30-mm Mk.44 Bushmaster II autocannon, a machine gun, or an anti-tank guided-missile launcher. Acting as loyal wingmen, M5s could fight alongside conventional manned tanks. Imagine an M5-equipped tank platoon, configured with four robotic M5s and one manned tank, and commanded by the human tank commander using voice commands just as they would command human-operated systems. The robotic tanks could reinforce combined-arms operations and provide striking power without risking friendly personnel.

The future of the MBT is both human and robotic. General Dynamics Land Systems introduced the newest US concept in tank design in 2022 with a new manned-tank demonstrator. AbramsX[9] is a 60-ton MBT with a hybrid electric engine and the computing power to operate a swarm of robotic wingmen. Three crew members operate the tank from inside the hull, providing for extra crew protection. Like the Russian T14 Armata,[10] the AbramsX has a robotic turret with a 120-mm automatic-loading cannon and machine guns to provide the firepower. AbramsX is designed to enable robotic wingmen and could also provide see-through armor for the crew to see both near and far. If embedded with appropriate edge-computing systems,[11] the tank could connect with every intelligent weapon system within range, allowing the crew to visualize a holistic view of the battlespace in real-time, and do all this without increasing the cognitive load on the vehicle commander. The military that develops and fields such a tank force, will generate a new level of mobile striking power that will play a central role in dominating the ground domain of any battlespace.

Tomorrow's Sensors, Shooters, and Jammers

A thorough study of urban operations in the Russia–Ukraine War reveals that conducting a siege of large city—such as Mariupol, which took the Russians 86 days using 14,000 troops, with artillery and airpower, to defeat a defending force of roughly 4,500 Ukrainians—is difficult and involves destroying the city and starving the population, a course of action most Western nations would not accept.[12] The other option, to assault a defended city, is a confusing and bloody affair, such as we saw in Kyiv in February 2022, and devours military power at a ferocious rate. Few armies other than the Russian Army and the Chinese People's Liberation Army have the manpower or the intentions to wage urban combat in the old way. Winning the city fight in the 21st century requires new thinking, but most importantly it requires effective intelligence, surveillance, and reconnaissance (ISR) that can be used to locate targets for sensors, shooters, and jammers. These new tactics require that sensors also act as strikers, sense-and-strike systems work in swarms, and that the attacker gain not only air dominance, but also control the stratosphere.

ISR sensors work best in open terrain. Buildings and reinforced-concrete structures inhibit line-of-sight sensors in urban settings. Several new technologies will improve the capabilities of sensor systems to overcome this in urban settings. The traditional kill chain, using separate sensing and shooting systems, involves a time lag that allows the target to evade the strike. In each of the three wars mentioned earlier, the time lag in the kill chain decreased the effectiveness of kinetic strikes. The exception is the use of robotic systems that sense, rapidly characterize the target as legitimate with a human in the loop, and then immediately strike. Sense-and-strike systems shorten the kill chain to seconds.

Israeli technology companies excel at developing robotic military systems and state-of-the-art loitering munitions. The Israeli-made Harop and Orbiter loitering munitions, for example, played a critical role in the Second Nagorno-Karabakh War. An Israel-based international defense electronics company, Elbit Systems Ltd, is building upon this expertise by creating networked, autonomous robots to dominate the battlespace. Elbit's LegionX[13] system connects multidomain robotic sensors of all types into one networked swarm. According to Elbit, "LegionX is an autonomous networked combat solution based on robotic platforms and heterogeneous swarms ... that enables tactical superiority at all echelons, enhancing

A swarm of drones scans the Cassidy Range Complex at Fort Campbell in a scenario conducted on November 16, 2021, during the final field experiment for the OFFensive Swarm-Enabled Tactics (OFFSET) program. Researchers with the Defense Advanced Research Projects Agency designed OFFSET with the goal of allowing infantry units to use swarms of upwards of 250 drones to accomplish diverse mission objectives in urban environments. (US Army photo by Jerry Woller)

efficiency and transforming capabilities in multidomain warfare. LegionX provides an advantage in peer/near-peer adversary combat scenarios, enabling coordinated deployment of swarms of connected, heterogeneous autonomous platforms and payloads." The LegionX network creates "one-to-many" control—one operator controlling dozens of systems—of air and land robotic weapons. A Wireless Local Area Network (WiFi) is used to exchange voice, data, and streaming video. In areas where there is no WiFi, Elbit's broadband tactical data communications network can provide software-defined radio (SDR) networking through land or air systems. For added resiliency, the network covers all NATO mobile frequency bands and does not rely on the global-positioning system.

A key element of the LegionX concept is the LANIUS loitering munition built for urban combat. LANIUS[14] is a beyond-line-of-sight loitering munition with the ability to sense and strike autonomously. The system communicates to other connected systems in the network through Wi-Fi or SDR networks. Onboard computing power and advanced algorithm artificial intelligence (AI) helps LANIUS avoid collisions with other objects and conducts simultaneous localization and mapping of its environment. LANIUS is a short-range weapon with a seven-minute flight time that can carry lethal or non-lethal payloads, fly as fast as 20 meters/second, or hover in one place. The mini drones are launched from a larger, longer-range, drone mothership. To clear buildings, a dozen LANIUS mini drones, armed with high-explosive warheads, can autonomously launch from the mothership to search for, and destroy, targets. LANIUS is still in testing, but in future city fighting, soldiers will use drones like LANIUS as grenades were used in World War II room-clearing operations. The difference will be that these smart drones will provide real-time video of what is in the building as well as delivering explosive effects. They will both sense—providing ISR information to friendly forces—and strike the enemy.

Swarm ISR and Strike

Every combat action involves finding the enemy and then striking him. No matter how advanced the weapons, putting human warfighters in harm's way in a city fight is a recipe for casualties. As the fighting in Ukraine's urban areas demonstrates, a city represents a complex battlespace that is dangerous to navigate and difficult to conquer. Urban terrain offers the defending force concealment, hardened positions, and the opportunity to ambush in every building and along every road. In the ongoing Russia–Ukraine War, the Russians have simply demolished cities and towns with artillery, rocket, and missile fires **before** sending in their troops. New thinking is required to avoid this scenario and the US Defense Advanced Research Projects Agency (DARPA) is hard at work to provide a solution.

DARPA is determined to develop and leverage AI to enhance military robotic systems for urban ISR and combat operations. The Deputy Director of DARPA's Information Innovation Office, Dr. Matt Turek, announced in March 2021, at a Defense Readiness Workshop, that AI is essential to over 120 of DARPA's most important programs. Turek added that DARPA is developing an "Explainable AI" program, XAI, to enable "third-wave AI systems, where machines understand the context and environment in which they operate, and over time build underlying explanatory models that allow them to characterize real world phenomena."[15] This will create AI-enabled systems that learn their environment to perform a variety of missions. Third-wave AI allows computers to become capable partners, rather than just tools, with human warfighters. An example of human–machine partnering was demonstrated in February 2023 with DARPA's Air Combat Evolution program, which enabled an F-16 aircraft to operate independently with AI. The aircraft, renamed as the X-62A or VISTA (Variable In-flight Simulator Test Aircraft),[16] flew several flights under AI control. Such AI will empower an unmanned aircraft to fly as a "loyal wingman" for manned aircraft. When used with loitering munitions, the AI will enhance autonomous and collaborative drone swarming. These tests have placed networked autonomous drones as a top priority for US Air Force funding and development.

DARPA's OFFensive Swarm-Enabled Tactics (OFFSET)[17] program addresses the ISR and strike problem for the urban battle using drone swarms. The DARPA website states the OFFSET program "envisions future small-unit infantry forces using swarms comprising upwards of 250 unmanned aircraft systems … and/or unmanned ground systems … to accomplish diverse missions in complex urban environments. By leveraging and combining emerging technologies in swarm autonomy and human-swarm teaming, the program seeks to enable rapid development and deployment of breakthrough capabilities." The concept combines collaborative, networked unmanned aerial vehicle swarms and UGVs with soldiers to provide an unparalleled sense-and-strike ability for the urban fight. Swarms of drones will act as both sensors and shooters, isolate buildings or areas in the urban battlespace and conduct urban raids. Instead of swarming soldiers into a city and accepting the heavy human casualties that this would entail, future city fights will swarm with flying and rolling robotic systems. In short, the use of networked, autonomous unmanned systems, employed in swarms, will change the methods of war.

Dominating the Sky and Stratosphere above the City

Real-time situational awareness is a force multiplier in a city fight and ISR drones are a basic tool in every modern military force today. Inexpensive, disposable small UASs are available to anyone with a few thousand dollars. Nearly every nation manufactures drones, with the most expensive and capable systems manufactured by China, the US, Europe, Israel, Turkey, and Iran. China is a drone superpower as it manufactures 80 percent of the commercial drones sold worldwide, with most of these being small UASs. One Chinese company alone, DJI Sciences and Technologies in Shenzhen, produces 70 percent of the world's consumer drone market.[18] DJI drones are used with great effect in the Russia–Ukraine War, comprising most of the small UASs used by Ukraine's Aerorozvidka (aerial reconnaissance).[19] Aerorozvidka is a unit of Ukrainian Army drone operators who have all become expert military drone pilots but were drone hobbyists before the war.

Drones offer military capability at reduced cost. Small, US$2,000 quadcopter drones can be used to see city blocks and can maneuver inside buildings, but observing the city battlespace from high and medium altitude is also necessary to fight in cities. Systems operating from a higher altitude provide a means to unmask enemy forces that are not inside buildings or hiding underground. Manned aircraft can provide medium- and high-altitude ISR, but in high-threat environments this mission is accomplished by high-altitude, long-endurance (HALE) and medium-altitude, long-endurance (MALE) UASs. Medium- and high-altitude ISR, however, is not enough when contemplating urban combat in a large city or megalopolis. Persistent ISR is needed.[20] To provide persistent ISR, a multilayered strategy is required that

Ukraine is using drones for intelligence, surveillance, reconnaissance, and strike missions to defeat the Russian invaders. In this photo, an R18 drone conducts a bombing test at the "Shyrokyi Lan" range in 2020. (Aerorozvidka drone unit)

includes a space layer that uses satellites, a stratosphere layer, and a mid-to-high atmosphere layer. Satellites in orbit around the earth, either in geostationary orbits or low-earth orbit, provide ISR from space. Manned aircraft and MALE and HALE unmanned systems provide atmospheric coverage. The gap appears to be in the second layer, the stratosphere, that extends from 14.5 kilometers (9 miles) to 50 kilometers (31 miles) above ground level.

To breach this gap, the US Army has been experimenting with stratospheric ISR unmanned systems. Craft operating from the stratosphere can take high-resolution images, transmit and relay communications with reduced latency, accelerate video feed and data processing, provide early warning of enemy threats, and jam an enemy's radar and communications systems, better than satellites in space. These capabilities will be essential for combat operations in a major city or a megalopolis. In 2021, the US military emphasized their interest in the stratosphere when US Central Command and the US Navy's Surface Warfare center published a Request for Solutions using stratospheric balloons and solar UASs.[21] Tests conducted in the past five years have focused on operationalizing the stratosphere for persistent operations in non-permissive environments.

One effort to gain a foothold in the stratosphere is the development of a stratospheric high-altitude unmanned aircraft call the Zephyr, manufactured by Europe's Airbus, and designed by QinetiQ, a UK Ministry of Defence-inspired company. Airbus calls the Zephyr a Solar High-Altitude Pseudo Satellite that can launch on demand from almost anywhere. Zephyr8 is one of the latest models of this ultra-lightweight carbon-fiber unmanned aircraft being tested by the US Army. It weighs less than 75 kilograms (165 pounds) and has a wingspan of up to 25 meters (82 feet). The wings and tail surfaces are large solar panels that power the aircraft for daytime operation and charge the lithium-sulfur batteries for nighttime operation. The Zephyr is so light that six-to-eight people can carry it for launching as its two propeller-driven engines lift it into the air. The US Army's Zephyr8 prototype flew over the southern United States, the Gulf of Mexico, and South America, at 60,000 feet, for 64 days in the summer of 2022 until it crashed in the Arizona desert

The US military has used aerostats for many years to provide real-time, persistent intelligence, surveillance, and reconnaissance information. In this November 13, 2019, photo, the 84th Radar Evaluation Squadron conducted an analysis and optimization of the Tethered Aerostat Radar System to support the Department of Homeland Security, and Customs and Border Protection, at Fort Huachuca, Arizona. The aerostat-borne surveillance system provided radar detection and monitoring of low-altitude aircraft and surface vessels along the US–Mexico border and select Caribbean areas. (US Air Force photo by GS-11 Deb Henley)

on August 8, 2022.²² The US Army did not reveal the precise reason for the crash other than the Zephyr experienced "events that led to its unexpected termination." Australia also purchased a Zephyr, but it crashed on September 28, 2019, when it ascended to 8,000 feet, executed a series of uncontrolled turns, was disabled by air turbulence, spiraled downward, and broke up during descent. The official reason for the crash was determined to be unstable atmospheric conditions. From January 28 to February 4, 2023, the infamous Chinese spy balloon that traversed much of the US—before it was shot down off the coast of South Carolina—stunned the world. This incident emphasized the military use of the stratosphere. Future urban operations will include not just dominating the air with traditional air superiority but dominating the stratosphere.²³

Placing sensors and jammers in the stratosphere is a growing military requirement. In most military operations a "HiLo" mix, combining expensive high-end systems with less expensive but capable low-end systems provides a battle-winning balance. Balloons can offer a "HiLo" mix solution for urban operations and can be configured to support swarms of unmanned systems with WiFi connectivity, to network the

The RQ-4 Global Hawk is a high-altitude, long-endurance remotely piloted aircraft with an integrated sensor suite that provides global all-weather, day or night intelligence, surveillance, and reconnaissance (ISR) capability. Global Hawks would be a vital part of the ISR effort in any future city fight involving the US military. The US Air Force expects to replace its Global Hawk fleet in 2027 with a more modern and capable system as the Block 40 Global Hawk fleet is no longer survivable against modern air defense. In this photo, an RQ-4 is towed across the flight line at Grand Forks Air Force Base, North Dakota, on October 23, 2020. (US Air Force photo by Senior Airman Elora J. McCutcheon)

drones in the urban canyon dead spaces. Such a "HiLo" mix provides redundancy, resiliency, and is less expensive than flying manned or HALE and MALE unmanned systems. The latest lighter-than-air (LTA) systems can carry large and sophisticated ISR and communications packages to provide persistent surveillance, network connectivity, and electronic warfare support over a city. Military LTA craft, called aerostats, can operate in the stratosphere to provide ISR for counter-unmanned aerial systems defense.

Aerostats are state-of-the-art LTAs that are tethered or free flying. These high-flying balloons operate above the altitude of aircraft but below the altitude of satellites, from 60,000 feet to 100,000 feet above the ground. One of the leading defense firms involved in military aerostats is the American defense corporation Lockheed Martin. The company was involved in military balloons with the US Navy before World War II and its latest, high-tech models are not your grandfather's blimp. The US used tethered tactical aerostats, at lower altitudes, for surveillance along the southern US border to combat drug trafficking. Since 2013, the Lockheed Martin 420K Aerostat System was the only ISR and communications balloon in daily use in the US until the Biden administration cut off funding and decided to ground them in late 2022. Another Lockheed Martin tactical aerostat model, specifically designed for military persistent surveillance and communications in lower altitudes, is the Lockheed Martin 74K Aerostat. The craft is 35 meters (115 feet) long and is tethered with a fiber-optic transmission cable. It can carry a payload of 500 kilograms (1,102 pounds).[24]

Higher-flying airships can provide significant wide-area surveillance and communications advantages for urban combat operations. Lockheed Martin's High-Altitude Airship (HAA) can operate in the stratosphere and provides the ability for unmanned persistent and sustained geostationary ISR, electronic warfare, and communications over a city. Because of its altitude, it is impossible to shoot down with most short-range air-defense systems and difficult to destroy with many longer-range systems. No tether is required as the HAA can maneuver in the airspace as directed from a ground station or satellite relay. The usual payload for aerostats includes a surveillance radar, inertial-navigation system, thermal-imaging cameras, electro-optical sensors, electronic-intelligence systems, and communication-intelligence systems.

Although the US has not armed its balloons, and has no intention to do so, other nations may not be as hesitant. Armed aerostats operating in the stratosphere could become geostationary weapons platforms to conduct precision bombardment.

ISR and Strike for the Next Urban Battle

The world is getting more dangerous. The likelihood of a great power war appears to be rising, with the possibility of multiple major wars breaking out at the same time. Fighting in towns and cities is the primary setting for combat in the current

Russia–Ukraine War and the primary lesson from this fighting is that urban combat cannot be avoided. If China was to invade Taiwan, most of the combat would occur in an urban battlespace. As much as we wish to avoid battles in cities, it is unlikely, and we must prepare, train, and equip to do so. Technology cannot address the challenge of urban combat alone, but it offers an alternative to the bloody fighting playing out in Ukraine.

In the near future, sense-and-strike drones, used in swarms, will dominate the airspace and stratosphere above a city. Platforms that provide for highly precise multi-domain ISR at safe ranges will improve the network, provide a means to jam enemy systems, and enable drone operations to win the city fight. AI, microminiaturization, and autonomous unmanned systems are driving these changes in warfare. In the next decade, military forces will transform from fighting as a network of independent capabilities, as we fight now, to a swarm of systems. In a 2014 study titled "Robotics on the Battlefield Part II: The Coming Swarms," author Paul Scharre predicted:

> Emerging robotic technologies will allow tomorrow's forces to fight as a swarm, with greater mass, coordination, intelligence and speed than today's networked forces. Low-cost uninhabited systems can be built in large numbers, "flooding the zone" and overwhelming enemy defenses by their sheer numbers. Networked, cooperative autonomous systems will be capable of true swarming–cooperative behavior among distributed elements that gives rise to a coherent, intelligent whole.[25]

In urban combat, these unmanned systems will provide the mass, reconnaissance, and strike abilities required to win an urban fight without excessive human casualties. As military forces deploy swarms of networked drones, drones will provide ubiquitous ISR and strike capabilities. Dominance in war will soon belong to swarms of networked robotic platforms that can sense, strike, and jam enemy forces autonomously. These weapons will not be inexpensive, but in the next war they will ignite a transformation as significant as the machine gun and the tank did in the 20th century. To adapt to these changing methods of warfare, we must think differently and act in time.

In May 2020, the US Army launched loitering munitions from a Black Hawk helicopter. Helicopters armed with systems such as the Raytheon Coyote Block 2 counter-unmanned aerial system loitering munition could provide mobile launch platforms, and command and control, to defend against enemy drones. (US Army photo)

CHAPTER THIRTEEN

The Big Blue Blanket—Light Tactical Aircraft for Counter Unmanned Aerial Systems Combat

> For the first time since the Korean War, we are operating without complete air superiority. As a result, USCENTCOM [US Central Command] has made the counter-UAS effort one of its top priorities and employs a variety of systems and tactics to defeat these threats.
> GENERAL KENNETH MCKENZIE JR., COMMANDER, US CENTRAL COMMAND[1]

It is late November and an American aircraft carrier sails in the contested waters of the Philippine Sea, surrounded by a number of escort ships. The sailors manning the aircraft carrier are on high alert, anticipating an enemy attack. As the sun rises on a beautiful, clear sky, Captain Tom Sprague orders the ship to face into the wind to launch aircraft. He wants his birds in the air this morning as the intelligence reports he received last night indicated a high probability of enemy action. A precision attack on his ship is his worst nightmare. "Alert, enemy inbound!" a voice from the ship's command center announces over an intercom heard on the bridge and throughout the entire ship. "Battle Stations! All hands Battle Stations!"

On the bridge, Sprague scans the horizon. In the distance, the air-defense systems on one of the escort destroyers engages the incoming targets. The antiaircraft systems fire furiously. The sky fills with explosions, but no hits. The two objects suddenly appear passing through the destroyer screen, approaching the carrier fast at low altitude.

The aircraft carrier's defenses blast away at the incoming enemy. It is all happening so fast. The enemy is flying straight towards them. "Brace for impact!" Sprague screams.

The first strike slams into the carrier's port air-defense positions. Fires and smoke billow from the port side. Sprague and his bridge crew are knocked to the deck of the bridge by the concussion. Ten seconds later, a second enemy strike pierces the flattop's flight deck and explodes in one of the hangars, setting a fire that burns the ship from stern (aft) to stem (bow).

***

Is this a scenario from a future naval battle against the Chinese Navy? No, it is the actual account of the USS Intrepid being attacked by two Japanese kamikaze aircraft on November 25, 1944, during an engagement off the coast of Formosa, now known as Taiwan. Though severely damaged, the ship managed to stay afloat. Its aircraft flew off to other carriers nearby. Casualties were 69 men dead and 35 seriously wounded. The carrier limped back to port and would be out of action for several months for repairs.[2]

The US Navy realized precision strikes by kamikazes required new tactics. Commander "Jimmy" Thach, who had earned a reputation for developing innovative fighter tactics to deal with the faster and more maneuverable Japanese Zero fighter in the early years of the war, came up with new air tactics to protect the fleet. The new tactic was dubbed the "Big Blue Blanket."[3] The name was derived from the blue paint job of the Hellcat and Corsair fighters that flew the Big Blue Blanket mission. This consisted of a combat air patrol flown 50 or more miles away from the main fleet. These aircraft flew dawn to dusk to sweep the skies of kamikazes as the Japanese suicide pilots could not target the ship's accurately at night. Thach's idea did not catch every kamikaze, but it helped protect the fleet and downed many a Japanese plane before they could get in range of the fleet.

Today, it is time to resurrect the Big Blue Blanket to provide armed overwatch and counter-unmanned aerial system (CUAS) operations for US forces. The challenge is how to do this? There are many new counter-drone systems available, with various defeat mechanisms, from jamming the enemy drone with an electronic-warfare (EW) attack to physically knocking it out of the sky with rockets, explosives, machine guns, or ramming. Many of these systems are very expensive—best used to defend fixed sites—and are not fit to provide mobile CUAS protection.

Drones to kill drones are needed, but until the entire CUAS system can be integrated into a full mobile integrated air-defense system (IADS) that also defeats incoming missiles, the part that is missing is the human-in-the-loop who is in command. One solution may be manned, propeller-driven, light tactical aircraft.

In an era of accelerating technology, every well-funded air force wants to be armed with the most up-to-date and sophisticated weapons. The US Air Force (USAF) exemplifies this rule as it operates some of the most complex and expensive aircraft in the world. The proposition to include propeller-driven combat aircraft in the high-tech USAF seems absurd and totally out of place but is this wise when we consider the pressing threat of unmanned systems? When it comes to shooting down drones, providing CUAS support, the most advanced technological multi-role fighter jets are not always the best option.

High-end platforms usually correspond to high cost and high complexity. Expensive aircraft like the F-35 Lightning II, can bring less, not more, capability to the CUAS mission. Slower, cheaper, manned aircraft, armed with specially designed munitions to hunt and shoot down drones, may provide a simple, reliable, and

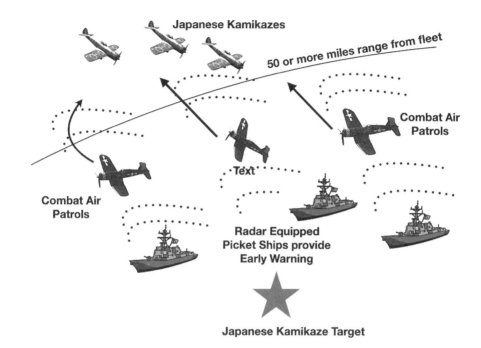

US Navy Commander John Thach developed the concept known as the "Big Blue Blanket," to provide the fleet protection against kamikaze attacks. Kamikazes were human-guided precision bombs. The strategy comprised sensors and shooters—radar picket ships stationed in a radius miles away from the main fleet, a daylight combat air patrol, and continuous flybys of the Japanese airfields. In the modern battlespace, robotic loitering munitions, cruise missiles, and long-range precision fires can strike targets with greater precision and accuracy than any World War II-era kamikaze. (Author's diagram)

effective CUAS solution. These aircraft could also act as nodes in an IADS system to knock out incoming missiles. In short, we may be able to use light tactical aircraft, what some call light attack aircraft, that would operate primarily behind friendly lines, to create a CUAS "Big Blue Blanket."

In a 2016 article in *War on the Rocks*, Major Mike Benitez, USAF, who was an F-15E Strike Eagle Weapons Systems Officer with over 250 combat missions, reported the cost of close air support was "over $64,000 per hour or $1,000 per minute per combat air patrol."[4] The cost of an F-15, for example, is about US$31.1 million, where the F-35, which is advertised as the ultimate stealth multi-role combat aircraft, is a staggering US$94 to $148 million per aircraft, depending on the model. The USAF's A-10 Thunderbolt II is possibly the best close-air-support jet engine aircraft; each one costs about US$46.3 million. It is also an airframe the USAF has tried to get rid of for years, only to have Congress step in to save it from retirement several times. The result of buying expensive exquisite platforms is that the Air Force does not have the budget to purchase enough aircraft to accomplish lower-priority close-air-support or CUAS missions. Is there a combat role today for low-cost, simpler to fly and maintain, propeller-driven aircraft?

Today, the US Army maintains a light-aircraft fleet of about 339 planes for EW operations and other missions. The E-O5, RC-12 and B-300 are Army EW aircraft with propellers and perform as special electronic-mission aircraft. There are three compelling reasons to consider turboprop aircraft in the modern battlespace: for economical and effective pilot training; for close air support and CUAS in low-intensity wars; and for the versatility they could provide as CUAS in higher-end conflict. NATO air forces are struggling to keep airplanes in the air and provide flying hours for pilot training. In 2017, the French and the British could only fly one-in-three of their multi-role jet combat aircraft. The rest were unavailable due to maintenance, repair, or lack of spare parts. The USAF aircraft-availability ratio is also dwindling due to expensive maintenance costs and increasing complexity. If you assess the problem from a ground combined-arms perspective, the need for light attack aircraft to perform close air support during counterinsurgency and counterterrorist operations, and CUAS in peer conflict, seems self-evident.

As NATO air forces rely on more expensive fourth- and fifth-generation fighter fleets for high-end missions, there are few assets available to work in the close-air-support role. Someday, unmanned aerial vehicles (UAV) may fill this role entirely, but high-end UAVs currently lack the versatility a manned light attack aircraft can provide. The USAF has toyed with the idea of purchasing a fleet of 300 light tactical aircraft, primarily to provide close air support for counterinsurgency operations, and the US Senate allocated $210 million for the purchase of light attack aircraft in the 2020 National Defense Authorization Act, but the USAF decided in February 2020 to cancel their Light Tactical Aircraft program as inappropriate for near-peer combat operations. Almost immediately after the USAF cancellation, US Special Operations Command (SOCOM) announced on August 1, 2022, the selection of the Air Tractor–L3Harris AT-802U Sky Warden for the "Armed Overwatch" program. SOCOM plans to secure around 75 manned, fixed-wing aircraft for close air support, armed reconnaissance, strike coordination and reconnaissance, and airborne forward air control.[5]

SOCOM did not mention the Sky Warden acting in a CUAS role, but there is no reason armed overwatch could not include CUAS. Laser-guided missile systems, such as BAE System's Advanced Precision Kill Weapons System,[6] provide an economical and immediate solution for the Sky Warden in the CUAS role. The tactical solution to provide responsive and effective CUAS with aircraft such as the Sky Warden is something US and NATO members should seriously consider.

Although the USAF believes turboprop aircraft might be acceptable for fighting terrorists, it is very skeptical about using turboprop light attack aircraft for close air support and CUAS missions in a peer fight in a sophisticated air-defense threat environment. This makes sense, if you think about close air support in the traditional way, but what if you could change the way it and CUAS are delivered? These aircraft could launch missiles or loitering munitions far behind friendly lines and out of the range of many enemy air-defense systems. For the CUAS mission, firing expensive surface-to-air missiles at drones is not cost effective. Modern, German-made InfraRed Imaging System Tail/Thrust Vector-Controlled air-to-air or surface-to-air missiles, which have been supplied to Ukraine to knock down Russian missiles and drones, cost roughly US$430,000 per missile, which is 20 times more than the average Iranian-made Shahed 136 drone, which costs about US$20,000. The Russians have used hundreds of Shahed 136 drones to pummel Ukrainian infrastructure targets and Ukraine has been using expensive rockets and missiles to knock them down. Using light tactical air, armed with cheaper laser-guided missiles, for the CUAS mission makes military and economic sense.

In the next decade, as the range, accuracy, and "intelligence" of air-launched missiles continues to improve, turboprop aircraft may offer an affordable high-low mix strategy. With the ability to launch intelligent, networked missiles from great distance, such as Rafael's Spike ER2, a fifth-generation extended-range missile, the way close air support is delivered could change. Today, the Spike ER2 can be launched from an air platform to a grid coordinate up to 16 kilometers (10 miles) away. Imagine if, in the near future, every missile is equipped with sensors that

There is a dangerous gap in counter-unmanned aerial systems (CUAS) defense. US Special Operations Command recently acquired the AT-802U Sky Warden armed overwatch platform to provide surveillance, close air support, and precision strike in the austere and permissive environments US special operations forces routinely operate. Why not use light tactical aviation for CUAS as well? (Air Tractor and L3 Harris)

continually sense and send information to provide the location of targets uncovered along their flight path? As the missile is on its way, the weapons officer aboard the aircraft, employing the missile's television-like non-line-of-sight engagement capabilities, sees the missile target area and can make the final changes in the missile's course to guide the weapon onto the target. As missiles like the Spike ER2 become common, and improve in range, accuracy, and sensor capability, the most important factor for the evolution of close air support may be for manned and unmanned aircraft to fly to a "firing box" and launch missiles from cover or "over the horizon." This high-low mix concept is a strategy to solve a complex problem by blending simple, proven, reliable, and affordable low-tech capabilities against a technologically advanced threat.

In August 2022, as already mentioned, SOCOM reviewed available light tactical piston-driven aircraft and selected the AT-802U Sky Warden for its armed overwatch program. SOCOM will spend US$175 million for the 75 aircraft. Although SOCOM's concept is to use these aircraft in irregular war operations, and they have not yet been specifically considered for the CUAS role, it seems only a matter of time, as the proliferation of drones of all types and capabilities increases with armies around the globe, that the concept of armed overwatch will include knocking down enemy UAVs.

The Sky Warden could be purchased by the Army in sufficient numbers to provide forward-deployed combat units with a "Big Blue Blanket" of armed overwatch for CUAS. In addition, our allies and friends would benefit from a mobile air platform that could provide CUAS and coordinate the CUAS effort. For CUAS to be effective, someone must be in charge of the CUAS effort and have the systems to win the fight. Creating an armed overwatch CUAS force with a commander in charge of a number of multi-mission aircraft and a package of aerial counter-drones would provide both the means and leadership for the CUAS mission. Imagine also if Ukraine had a fleet of reliable and versatile light tactical aircraft to provide mobile armed overwatch CUAS for the defense of cities such as Kyiv and Kharkiv? The US Army does not have this capability today and, as we can learn from recent wars, effective CUAS is vital to restoring maneuver in the battlespace.

Another option for the US Army is to designate the CUAS mission to Army Aviation units. If helicopters were refitted with new sensors and armed with rockets such as the AGR-20A Advanced Precision Kill Weapon System II (APKWS II) 70mm/2.75 inch laser-guided rocket, they could defeat a variety of UAS threats. Winning the CUAS fight will require a commander in control of the appropriate combat forces. Army Aviation, commanding the CUAS fight with appropriately armed helicopter formations, could provide this to generate a Big Blue Blanket (or an Army Big Green Blanket) over US and allied ground forces.

Throughout the history of war, military forces have waged combat with a mix of weaponry, some revolutionary, some new, and a lot of old kit. In most cases, simple and reliable wins over complex and high maintenance. With only a few expensive jet aircraft available to provide close air support, most of these high-end aircraft are incapable of flying slow enough to target drones. Expensive and complex attack helicopters are unable to provide the extended loitering time that fixed-winged aircraft can provide. The option of a simpler turboprop aircraft for launching precision missiles and for CUAS armed-overwatch operations can provide a versatile multi-mission option that can operate outside of the enemy's man-portable air-defense systems' envelope. Until a fully robotic solution is developed—that is mobile, versatile, inexpensive, and effective—the use of manned light tactical aircraft should be considered. A mobile CUAS "Big Blue Blanket" based upon light tactical aircraft could be a vital part of defeating enemy drones today. The essential lesson is that the CUAS fight must be a priority.

Textron's M5 Ripsaw Robotic Combat Vehicle (RCV) is designed to provide unmanned combat power for the emerging hybrid human–robotic force. Command versus control of RCVs is the key next step, as it is for flying drones. These unmanned tanks might soon be commanded by human tank commanders as a "wingman tank" rather than controlled by operators. See https://www.textronsystems.com/products/ripsaw-m5#overview. (Textron Systems)

CHAPTER FOURTEEN

Developing a Hybrid Human–Robotic Force

> We may be on the leading edge of a new age of tactics. Call it the "age of robotics." Unpeopled air, surface, and subsurface vehicles have a brilliant, if disconcerting, future in warfare.
>
> CAPTAIN WAYNE P. HUGHES JR., US NAVY (RETIRED)[1]

There are dust clouds on the horizon. A newly formed American force composed of Robotic Combat Vehicles-Heavy (RCV-H) are in position, screening a sector on NATO's eastern flank. The robots weigh 30 tons, carry a lethal automatic cannon and several anti-tank guided missiles (ATGMs), and are lighter and smaller than M1A2 SEPv3 Abrams tanks. The robots are deployed along the edge of a forest. They quietly scan the terrain with steady thermal and infrared "eyes," searching for the lead elements of the advancing enemy force. In addition to the ground systems, the US commander has unmanned aerial systems (UAS) hovering overhead, providing additional situational awareness.

The RCV–Hs are stealthy and specifically designed with reduced thermal and electronic signatures that make them nearly invisible to the enemy's sensors. A kilometer to the rear of the RCVs are a group of optionally manned fighting vehicles (OMFV), next-generation combat vehicles designed to control and integrate robotic combat systems. Inside each OMFV is a bank of computers, each manned by an operator who controls an RCV–H. This human-in-the-loop system enhances the RCV's state-of-the-art sophisticated AI that enables the vehicle to move across rugged terrain and avoid most obstacles.

Radio traffic inside the OMFVs increases as unmanned aerial vehicles (UAV) report the enemy moving into range. The OMFVs are nestled into protected reverse-slope positions. The controllers are confident. They have practiced this fight in a hundred simulations and each time their RCV–Hs have exacted a heavy toll from the enemy. Guided by other networked UASs flying far forward, the plan calls for long-range precision rocket and artillery fires to hit the advancing enemy.

As the enemy moves closer, the RCV–Hs will launch ATGMs and then laser-designate the enemy vehicles for further precision strikes by attack helicopters and unmanned combat aerial vehicles (UCAVs). If any of the attackers make it through this precision-fire hell, the RCVs will open up with direct cannon fire. As the enemy attacks, consistently monitored and targeted along the way, his destruction is all but guaranteed by precision fires. After all, it is a simple matter of mathematics. The controllers stare at their screens, slowly moving their RCV–Hs into position and preparing to give the order to fire.

The interiors of the OMFVs go dark. Artillery fire slams into the ground nearby, like a blow from a gigantic sledgehammer. The roar of the explosions is deafening, even inside the armored protection of the OMFV. The lights quickly come back on, but to the controllers' shock they soon find themselves with no communications to their RCVs or UAVs. Enemy cyber electromagnetic activities have jammed communications. The enemy strike zone is inundated with electronic-warfare (EW) jamming and pummeled with rocket and artillery fires.

As the shells fall, ripping off antennas and adding further to the chaos, the controllers look at each other and wonder what to do. The enemy's outmoded, human-operated, T-90MS main battle tanks (MBT) will soon overrun the RCV–H positions. These robots do not fight without controllers. With no alternative, the OMFVs retreat, hoping they can outrun the onrushing T-90MS. In the next war, it is clear that the side who controls the electrons—and denies them to the enemy—wins.

***

In 2020, US Army Yuma Proving Ground tested the Advanced Running Gear for potential use on the future Optionally Manned Fighting Vehicle (OMFV). The OMFV will replace the Bradley Fighting Vehicle and be capable of operating in manned and unmanned modes. OMFV is expected to be fielded in 2029. (US Army photo by Mark Schauer)

A dismal scenario, certainly, but one distinct possible scenario illustrating the fury and chaos of future wars and the use of human-on-the-loop (HOTL) robots. Today, robots are part of the arsenals of every advanced military force. The US Army is developing a family of RCVs and "plans to develop, in parallel, three complementary classes of Robotic Combat Vehicles … intended to accompany the OMFV into combat, both to protect the OMFV and provide additional fire support. For RCVs to be successfully developed, problems with autonomous ground navigation will need to be resolved and artificial intelligence must evolve to permit the RCVs to function as intended."[2] As Maj. Gen. Ross Coffman, US Army, said, "When you couple that with these multi-ton, payload-agnostic war machines that are silent and magically appear, these systems are the equivalent of the Ghosts of Patton's Army."[3]

In April 2018, former Secretary of the Army, Mark Esper, agreed to accelerate the OMFV timeline to equip initial units in the first quarter of FY26. On July 1, 2022, the Army announced a Request for Proposal for the third and fourth phases of the OMFV to open up competition for prototypes. The Army is moving forward with the integration of humans with robots, known as MUM-T (Manned Unmanned Teaming), and heralds this as the future of war. That may be so, but we might examine recent developments in MUM-T and determine where the experts may be wrong.

A Wicked Problem

Today, the exponential increase in speed of technological change continues to transform the battlespace. The desire to protect soldiers by using robotic weapons is strong and justified. Putting a robot in harm's way is better than risking a human life, but this is not a tame problem—one that is solved by applying the right equation or algorithm. Process does not solve a wicked problem. A wicked problem is a multidimensional challenge that may have no solution and that has no clear delineation between cause and effect. Wicked problems are often intractable and defy reduction to smaller, solvable problems. Wicked problems cannot be solved with a linear-management or critical-command approach.[4] To address a wicked problem, first ask the right questions rather than work to provide the right answers.

War is a wicked problem, as the thinking enemy "always gets a vote" and no plan survives first contact with the enemy. Nothing will ever turn war into a tame process. War is, and always will be, a wicked problem, resistant to solutions, for which there is no singular solution but only multiple possibilities, some less bad than others. Therefore, overwhelming mass can win battles to cover the unforeseen complexity that always develops. Superior weapons will also help and can sometimes be decisive. Developing military technology to win a future war, however, is also a wicked problem.

The duality of military technology—every weapon creates both strengths and weaknesses—is difficult to learn in an era of prototype warfare. If you get it wrong and invest significant time and resources, you can move from strength to catastrophe in rapid order. As soon as you make contact with the enemy, technological solutions only provide part of the solution.

Consider the French effort to build the Maginot Line as their best war-winning technology created by their most creative minds. The French believed they had learned from the carnage of World War I and intended to use this foresight to win the next war. They sought a prescription that would take out as many variables from combat as possible. The Maginot Line was researched, perfected, funded, built and manned, only to be bypassed, turned, and thus neutralized. The French attempted to transform war from a wicked problem into a tame one and failed. Their mistake was further magnified as their solution solved the problem of the previous war. It did not prepare the French Army for the next war. Imagine what might have been accomplished if the money and energy to create the Maginot Line was spent to forge the French Army into a hard, tenacious, and resilient combat force that was ready to win the next war.

Asking the Right Questions

One of the most important skills needed to solve a wicked problem, such as the use of RCVs in the future battlespace, is to start by asking the right questions. The questions you ask in any situation define the answer. The US Army has asked creative developers and agencies to determine how to control ground robots in the future multidomain battlespace. Scientists, researchers, and think tanks are working on this problem. Artificial intelligence (AI) that is working for driverless cars is being enhanced to provide RCV cross-country mobility, but the AI challenge of the battlefield is magnitudes more complex than driving on a road and, therefore, requires human control. The current vision is to control RCVs from a nearby OMFV, with a human operating each RCV or, as the technology progresses, multiple RCVs. A human-in-the-loop will control or oversee movement and solve the ethical issue of engaging targets with lethal force. The current focus is on **control** of the robot.

An aerial version of MUM-T was tested with success in recent combat in Iraq and Syria. AH-64E Apaches have integrated both the Textron RQ-7 Shadow and General Atomics MQ-1C Gray Eagle in order to transmit and receive real-time streaming video, and to control the UASs to engage targets far forward of the helicopter. The AH-64E is digitized, with the latest "level 4 man–unmanned teaming" systems. For helicopters, the US Army defines the levels of interoperability (LOI) for MUM-T as: LOI 1, indirect receipt of UAS payload data; LOI 2, direct receipt of UAS steaming video and other sensor information in the cockpit; LOI 3,

Manned Unmanned Teaming (MUM-T) is a US Army effort to combine soldiers, manned and unmanned air and ground vehicles, robotics, and sensors to increase situational understanding, resiliency, and lethality. (US Army image)

receipt of UAS video and pilot remote control of UAS sensors/payload; LOI 4, video sharing, sensor control, and manipulation of UAS flight path; and LOI 5, full UAS control from take-off through landing. This MUM-T capability reduces the Apaches' sensor-to-shooter lag, enabling faster and more accurate engagements.

MUM-T is a revolution in technology that is now combat tested for helicopters in a counterinsurgency battlespace, but how will it work for ground RCVs in a peer-on-peer conflict? Right now, the Russians have significant combat experience controlling Uran-9 RCVs in combat in Syria in 2018—with ambiguous results. This form of prototype warfare is still in its infancy, however, and the Russians are learning as they fight. There are no reports as of this writing of the Uran-9 in combat in Ukraine but, on January 18, 2023, *Newsweek* reported that "Dmitry Rogozin, the head of the 'Tsar Wolves' military advisor group in Russia, said on his Telegram page that the combat robots would be tested out in the eastern Donbas region and would fire upon enemy targets in the affected area with its own fire weapons."[5]

Command or Control the Robot?

The question most robotic-systems designers and military leaders are asking is "How best to control a ground RCV?" Is this the right question? Controlling puts soldiers in OMFVs as controllers, not fighters. When the RCVs go offline, are damaged or destroyed, the combat power of the formation shrinks to insignificance since the RCVs are designed to be controlled by human operators.

Controlling a ground RCV is more difficult than controlling a UAS. Unlike the sky, the ground is multidimensional, with many traps and snares to avoid—in short, another wicked problem. To avoid losing the expensive RCV, the controller will move slowly, careful not to roll the vehicle off into a ditch. With a view restricted to the narrow view of the robot, the controllers will move their RCVs at a snail's pace. Anyone who has ever played a first-person-shooter video game understands this: if you only had one life and had to give up the game if you "died," would you move fast or slow? Every controller, therefore, will hang back, trying to protect the RCV and generating a very defensive attitude.

To tackle this wicked problem, we must first understand the nature of war and then decide whether ground RCVs should be "commanded" by leaders or "controlled" by operators. Imagine if RCVs (with adequate but simple AI), reinforcing existing M1A2 SEPv3 Abrams tank platoons to thicken the force, were commanded by the platoon leader's voice commands. In this situation, the human tank commander takes up the slack in situational awareness, which the AI cannot cover, by maneuvering his own tank and commanding the RCVs (as a platoon leader

The M5 Ripsaw, developed by US companies Howe & Howe and Textron Systems, is the fifth generation of Ripsaw, providing speed, mobility, and unmanned capability. The M5 can silently maneuver and keep pace with current and future maneuver forces, pushing capabilities beyond the human formation. In May 2021, Textron Systems and Howe & Howe delivered the fourth M5 to the US Army in support of the Robotic Combat Vehicle-Medium program. (Textron Systems)

would give orders to a human-operated wingman tank). The AI required to react to the platoon leader's commands is relatively simple and could be transformed into a series of battle drills.

The communication link between the human commander and the RCV would use multiple means, such as a point-to-point millimeter-wave link between the two vehicles, minimizing the enemy's ability to jam commands to the RCV in combat. To reduce the cognitive load on the human tank commander, they do not control the RCV's every move but executes a set of simple digital or voice commands. For example, on command of the human tank commander, the RCV wingman would execute three battle drills: follow in column; move to line and engage targets in the area designated by a human commander; or move forward to a waypoint, scan the area ahead, and engage targets if ordered, firing at the target area designated by the human-controlled weapon. Here, there is no bank of controller-operators in an OMFV, just a tank commander issuing commands to his RCV wingman. The human commander fighting the battle is the most important part of the system.

Like the French and their Maginot Line, we have been trying to transform war into a tame process that can be controlled from a distance. As a result, we have been asking the wrong questions. You cannot solve wicked problems with a tame, linear mindset. You win wars with fighters, not controllers. There is little doubt the convergence of AI with robotics and autonomous systems will change the nature of warfare, but the critical issue today is to think about how to integrate them with our most capable system—the human soldier. The question of command versus control of robotic systems, therefore, is the fundamental factor in determining the effectiveness of robotic systems for the next war.

Forging battleshock in the next war demands well-trained leaders. In this May 18, 2015, photo, a leader in the 1st Armored Brigade Combat Team, 3rd Infantry Division, designates the location of the battalion's tactical operations center in the Hohenfels Training Area. (US Army photo by Sgt 1st Class Caleb Barrieau)

CHAPTER FIFTEEN

Command Post Rules

> The history of failure in war can almost be summed up in two words: Too late. Too late in comprehending the deadly purpose of a potential enemy. Too late in realizing the mortal danger. Too late in preparedness. Too late in uniting all possible forces for resistance.[1]
> GENERAL DOUGLAS MACARTHUR

Recent wars demonstrate that unmasked command posts (CPs) are at great risk. In their current configurations, our CPs are nearly impossible to mask and extremely difficult to defend. The majority of our configurations—most of our battalion, brigade and division CPs—will not survive the enemy's first strike. Destroying command and control is the essence of 21st century warfare. Finding and targeting your CP is at the top of the enemy's "To Do list." In the meantime, developing survivable CP configurations now, using existing equipment, must be a priority. Going forward, experimenting with new equipment, and determining what is needed, and demanding changes, is vital.

Imagine what it will take to reconstitute a destroyed brigade or division CP. This would require significant time and great effort and expense to train soldiers, non-commissioned officers, and officers to operate an effective CP. We cannot afford to lose CPs the way the Russians and Ukrainians have in the Russia–Ukraine War. Russian CP configurations at the beginning of the conflict were not much different than our own. Since the onset of hostilities, their losses have been horrific.[2] We must "rethink command posts for this new era of warfare. In the face of this immediate threat … command posts will need to adapt to such an extent that they will be unrecognizable to the generation of leaders that fought in Iraq and Afghanistan."[3]

To avoid turning CPs into graveyards, commanders must act now to adopt new tactics, techniques, and procedures (TTP) for CPs. As Gen. James McConville, Chief of Staff of the US Army, said in October 2022: "In the future, the battlefield will be so lethal, and there'll be the ability to gather [targeting] information on where our command posts are, so we're going to have to move them very, very quickly, and they'll have to be dispersed and smaller."[4] The following rules should inspire that effort.

1. **The CP's Purpose is to Enable Mission Command:** A CP has one purpose—to facilitate Mission Command to win the fight. Mission Command must be more than an idea, it must be ingrained in the unit's culture. Every leader must understand, teach, and live Mission Command. "Mission Command is the Army's approach to command and control that empowers subordinate decision making and decentralized execution appropriate to the situation."[5] Mission Command empowers leaders to act when the situation has changed, orders no longer apply, and communication with higher command is lost. The CP facilitates the commander's execution of Mission Command by aiding the commander to better understand, visualize, describe, direct, and evaluate combat operations.

For command groups and CPs to survive in the modern battlespace we must be excellent at Mission Command across the force. Leaders must focus on substance over form. Leaders need to be proactive, located forward at the point of decision, and acknowledge that there will always be some degree of risk and uncertainty. The influential military thinker J. F. C. Fuller explained the need for active and vigorous command when he wrote: "In the World War, nothing was more dreadful to witness than a chain of men starting with a battalion commander and ending with an army commander sitting in telephone boxes, improvised or actual, talking, talking, talking, in place of leading, leading, leading."[6]

2. **Develop Distributed Mission Command:** CPs must disperse to survive. "Distributed Mission Command" is an emerging concept, defined as the execution of Mission Command using smaller distributed command nodes to execute the functions of the command post without staff co-location. The goal of distributed mission command is to enhance continuity and survivability of the command function in a transparent and lethal battlespace. (Author's definition). The execution of Distributed Mission Command arranges the CP infrastructure into resilient "functional nodes" that are small, mobile, and dispersed throughout the battlespace, yet remain in constant communication.[7]

3. **There is no Sanctuary:** In today's battlespace, every CP can be located, targeted, and hit. Sensors will find your CP and long-range precision fires will strike it. A study of CP strikes in the Russia–Ukraine War vividly demonstrates this truth.[8] General Mark Milley, the US Chairman of the Joint Chiefs of Staff, said "The probability of being seen is very high. In a future battlefield, if you stay in one place for longer than two or three hours, you'll be dead." That statement is optimistic. There are no sanctuaries, no rear areas, and no safe areas. We must accept risk, but we can mitigate risk through proper masking and dispersion.

4. **Mask to Survive:** Masking is the full spectrum, multidomain effort to deceive enemy sensors and disrupt enemy targeting. We must mask in the areas of optical (be the best at camouflage), thermal (reduce heat signatures), electronic (lower emissions and manage electronic signatures), and acoustic (dampen sounds). Consider masking in all decisions. For instance, do not conduct rehearsals in open terrain.

Enemy sensors could be watching and, if they are, they will see and report on your plan and what you intend to do. Find a covered location or conduct the rehearsal inside a building.

Take every opportunity to deceive enemy sensors by using decoys. Decoys can be as simple as wooden silhouettes (optical masking), smudge pots (thermal masking), thermal targeting panels (thermal masking), or as sophisticated as remote radio transmitting (electronic masking—protecting the emissions of friendly communications and electronic systems against enemy electronic-warfare support measures), and cyber operations (fill the enemy's sensor monitors with dozens or hundreds of false-positive targets). Most targeting today is from overhead systems, so use decoys that provide overhead imagery views of false targets. Set up a false CP with a tent and camouflage net. Empty tents with fake antennas may look like a CP to a drone operator. If you do not have any fancy pre-made decoys, make your own.

Electronic masking is very difficult but must be considered and planned for:

> One way to mitigate the usage of our digital-aged "easy button" is to implement mandatory communications exercises prior to field immersion of the training environment. A unit's lack of comfort across the board with seamless shifting between the primary, alternate, contingent, and emergency (PACE) communication methods seems to stem from the lack of planning and practice of its PACE plan.[9]

Use radio-communication listening silence as your preferred mode of operation. Determine how to link distributed and dispersed CPs with wire or electronic cable to make it harder for the enemy to detect your electronic signature. Units should take "steps to camouflage their electromagnetic footprint similar to the effort placed on their visual signature. Simple mitigation techniques such as placing antennas on the side of a hill to provide maximum exposure to friendly forces but limit line-of-sight to the enemy will cut down on errant electromagnetic signatures."[10] Develop and enforce the discipline and TTP to turn off all but essential electronic systems for periods of time. As crazy as this sounds in our constantly connected world, CPs that can go dark, on command, and restart according to plan, have developed an extraordinary defensive measure.

Do not allow soldiers to carry mobile phones in the battlespace as they are extremely easy to locate and target. Remember the Russians that were targeted and killed at Makiivka on January 1, 2023, due to their uncontrolled use of mobile phones.[11]

Mask logistic support operations for your CP. All CPs require frequent resupply of all categories of supply. Logistics vehicles arriving at a masked CP location reveal your position to the enemy. Enemy UAVs and sensors will follow support vehicles to the resupply location to locate your position and target you. Consider resupply at night by drone, speed-bag drops, or move to pre-designated supply caches, rather than have resupply come to you.

5. Train as You Fight: "At its core, the current command-and-control dilemma reflects an imbalance in the functional requirements for command posts to be both effective and survivable."[12] To redress this imbalance, CPs must not only be redesigned and reconfigured but, most importantly, CP teams must be expertly trained. You do not rise to the level of your expectations; you fall to the level of your training. We must train CP crews to perfection with battle drills, just as you would train a professional sports team. If your CP team does not have CP battle drills, such as "Rapidly locate and occupy a CP site," "Mask a CP site," "Take actions upon an enemy drone attack," or "Relocate a CP to another designated location," then develop them, document them in a unit Standard Operating Procedure (SOP), then continually train to exceed the standard.

6. Go Micro: Go micro in all things and avoid the macro. Bigger is not better. Bigger will get you killed. Stop concentrating all your brainpower in one place. A small footprint is the goal. Modern war is a shot to the head and CPs are at the top of the enemy's targeting list. Stop erecting "Taj Mahal TOCs" that are easy to find. Do not set up huge tactical-operations centers in tents. Tents offer no protection

This image was taken at the National Training Center by the author in 2022. Armored Brigade Combat Teams now have M1087 Expandable Van Shelters, known as "Expando-Vans," to use for command posts (CPs). These vans are easier to set up than tents but offer no significant armor protection. A brigade CP in a van like this is very difficult to mask and defend, and will be located and targeted. (Author's photo)

from drones or artillery strikes and "expando-vans,"[13] such as the newly designed M1087 Expandable Van Shelter, are not much better. Although the expando-vans provide an effective CP set-up, they are not mobile CPs as you cannot operate safely from inside them when they are on the move. These shelters are relocatable, meaning they take time to set up, break down, and relocate. In the event of an attack, they provide little protection. Seek hardened areas. Never position a CP in an open field when a gully, forest, town, or city is nearby. Ditch the big computer screens for smaller communications surfaces. Develop a mesh CP configuration and network your CP nodes. This will reduce the electromagnetic signature of CPs and increase their survivability. Go micro.

7. Red Team Your CP Configuration: Red teaming applies a devil's advocate approach to your plans and actions. Red teaming is defined as a "structured, iterative process executed by trained, educated and practiced team members that provides commanders an independent capability to challenge plans, operations, concepts, organizations and capabilities in the context of the operational environment and from our partners and adversaries' perspectives."[14] Always red team your CP placement and configuration by having someone play the enemy to analyze your CP's weaknesses. For example, after a CP location has been selected, but before it is occupied, designate someone to red team the proposed location to determine the enemy's ability to find and target the CP. Adjust the location and configuration as warranted after reviewing the red-team analysis.

8. Organize in Threes or Fours (but no more than four): The target that sticks out, gets hammered. Employ no more than 3–4 vehicles together in one location for all echelons of command—from battalion to corps. Develop a mesh CP configuration. A mesh Command Post (CP) configuration is the distribution of the CP infrastructure into resilient "functional nodes" that are small, armored, mobile, dispersed, and masked throughout the battlespace, yet remain in effective communication, with each other in the execution of mission command and are ready to assume command when required. Each node becomes a means to mitigate the risk of the nullification of other nodes. The ideal mesh CP configuration is a flexible self-forming, self-healing, and eventually self-organizing tactical network arrangement of command nodes. (Author's definition.) This will seem insane to a generation of leaders who fought in Afghanistan and Iraq, but it must be done, or we must become expert at reconstituting destroyed CPs, which is not a very viable alternative. Each 2–3-vehicle node in the mesh is connected to data, using tactical cloud computing or other means, but dispersed for protection from enemy fires. Three to four vehicles look like a platoon to most enemy sensors. The battlespace will be full of platoons. A dozen or more vehicles is a target that sticks out, looks like a CP, and offers a tempting target.

To mask, we must generate "false-positive" signals to deceive enemy sensors by avoiding obvious concentrations of vehicles. Network these groups of 3–4 vehicles with each other to form a mesh CP structure and generate "Distributed Mission Command," with each node dispersed from the other by hundreds of meters or more (depending on terrain and communication links). For example, a mesh battalion-level CP configuration could consist of three distributed nodes, dispersed according to the terrain and threat, all operating from available armored vehicles and all using the same ADCOP:

> Node 1: Commander with the Fire-Support Officer and two supporting staff assistants;
> Node 2: Operations Officer (S3) with 2–3 supporting staff assistants;
> Node 3: Intelligence Officer (S2) with 2–3 supporting staff assistants.

Each node in this mesh CP configuration should be capable of taking over the fight, in succession (Commander, S3, S2), if one of the nodes is put out of action. Again, those who say this cannot be done should consider how difficult it will be to reestablish command and control after the battalion, brigade, division, or corps CP is destroyed. The Armenians, Russians, and Ukrainians have learned about CP survivability the hard way. By studying their CP failures, we can gain foresight, think differently, and start training and equipping to meet the threat.

9. Armor Up: In the Russia–Ukraine War, 90 percent of the casualties are caused by artillery fragments. Maximize available armored vehicles for protection from enemy fires and to conduct Mission Command on the move. Each armored vehicle node should be equipped with the appropriate communications and computer systems and be capable of taking over the fight independently. Use whatever armored systems you have. Any armored vehicle is better than a tent or a soft-skinned, wheeled vehicle. Tracked vehicles are better than wheeled vehicles in most cases and a quick study of the effects of mud in Ukraine verify this claim. In 1982, Gen. Donn A. Starry said, "… experience convinced us that the corps battle cannot be fought from the Main CP and we believe the evidence is sufficiently compelling for us to field an armor-protected TAC CP [tactical command post] with sufficient equipment and personnel to track the battle and issue timely orders."[15] His insight was true then and is more poignant today in a transparent battlespace against a peer opponent. Light infantry will need to either add armored vehicles, occupy hardened urban facilities, or dig in deep.

10. Operate and Practice Mission Command on the Move: Moving targets are harder to hit. Do not stay in any position for an extended time. Mission Command on the move allows commanders to lead from the front with a smaller tactical footprint and with a better chance of survivability. The US Army is working on a Mounted Family of Computer Systems to support Mission Command on the move but fielding this will take time. In the interim, units must organize and equip

… COMMAND POST RULES • 175

Command Post Survivability Index

Review your CP and select your level of each category based on your analysis of the situation and your set-up.

LOW ⟶ HIGH

Category	1	2	3	4	5	6	7	8	9	10	Number	Color
Optical Masking (Camouflage)	1	2	3	4	5	6	7	8	9	10		
Thermal Masking	1	2	3	4	5	6	7	8	9	10		
Electronic Masking	1	2	3	4	5	6	7	8	9	10		
Acoustic Masking	1	2	3	4	5	6	7	8	9	10		
Dispersion	1	2	3	4	5	6	7	8	9	10		
Mobility	1	2	3	4	5	6	7	8	9	10		
Active Protection (CUAS & AIR DEFENSE)	1	2	3	4	5	6	7	8	9	10		
Armor Protection (Is your CP in an armored vehicle?)	1	2	3	4	5	6	7	8	9	10		
Occupy inside a City or Town	1	2	3	4	5	6	7	8	9	10		
Enemy's Kill Chain (Do you have a plan to disrupt it?)	1	2	3	4	5	6	7	8	9	10		
TOTAL												

COLOR CODE 1-2 = Black   3-5 = Red   6-7 = Amber   8-10 = Green

This form is a simple way to rate the survivability of a command post (CP). Review the situation and your CP setup and decide the survivability rating based upon your best judgment. Add up the score. Reflect and think of ways to improve the CP's survivability. (Author's diagram)

existing armored vehicles and command-and-control equipment for this purpose. Units must practice Mission Command on the move.[16] The US Army's I Corps is developing a Distributed Mission Command setup that executes on the move. "In a recent experiment in Guam using four Strykers loaded with advanced communication capabilities, the I Corps worked to prove that it can pass important battlefield data, including fires and targeting information, between platforms, even while they are in transit in the air or on a boat."[17]

11. Each CP Node must be Fully Capable of Assuming Command when Required: Train every CP node to assume command in case the commander is killed or out of communication. In combat, disruptions in command can be devastating. Any node must be ready and able to assume command. Have a node succession-of-command plan and practice this in training. Develop redundant command and communications solutions, thereby avoiding a single point of failure.

12. Occupy Urban Areas: Never set up in open fields. Find a gully, wadi, or forested area to position your CP. Reverse-slope positions are no longer safe, as the enemy can attack from above, but they are better than forward-slope positions. Operate CPs from towns and cities. Built-up areas make it harder for the enemy to target, provide false positives, and offer opportunities to implement decoys. A decoy CP in a building with fake antennas sticking out all over, may draw enemy fire and save your CP. Transmitting from a building at the other end of town, using a displaced transmitter, might fool the enemy into thinking your CP is somewhere else. Choosing the right building or basement may offer better physical protection. Buildings can provide cover in the form of reinforced steel and concrete. Consider the civilian population and act accordingly. If the civilian population is friendly, work with the civil authorities and evacuate civilians from sensitive areas. Develop, rehearse, and practice the TTP for your CP to operate from inside a town or city. Do this now, rather than after the shooting starts.

13. See Your CP as the Enemy Sees You: Know what the CP looks like from above. Procure several quadcopter small unmanned aerial systems (UAS) from the US Government Service Agency (GSA) catalog with unit funds and use these every time you conduct combat training. Fly these small UASs overhead and take pictures and videos of your CP. Use these images and film clips for detailed, visual, after-action reviews on your CP's survivability and adjust your setup, training, TTP, and SOP accordingly.

14. "What if" Drills: Develop a "What if the enemy hits me with fires in this location?" plan for every CP location. Fires include long-range precision fires, artillery, and UASs. Designate a leader to act as a red team to help generate two or three "what if" scenarios to wargame your CP location and setup.

15. Obscuration: Develop plans to use smoke to obscure your position. When attacked, employ every form of obscuration, including smoke, preferably

At the US Army's National Training Center, Fort Irwin, California, the opposing forces regiment, known as the OPFOR, practices and perfects its Mission Command skills nearly every day of the year. This frequency of training makes them an elite combat unit. This photo displays the OPFOR's Common Operational Picture, digital and analog, used to fight and track the battle. (Author's photo)

multi-spectral smoke. Have a jump-plan to move your CP to a designated rally point if attacked. Russian forces are using TDA-2K smoke vehicles and other smoke generators on the move, billowing multi-spectral smoke that hides positions from infrared sensors. Learn from this.[18]

16. **Create an ADCOP:** Develop an All-Domain Common Operational Picture that facilitates Mission Command. Developing a common operational picture, that is understood by all leaders in a command, is critical to Mission Command. Use modern digital systems along with legacy map backup as a contingency if the CP loses power. Operate Mission Command as a service, whereby every CP node has an updated ADCOP, and any commander could fight the battle from any existing CP. As the US Army I Corps Chief of Staff emphasized: "If we're running the Corps and performing our command-and-control functions from three, five, six locations as opposed to just one, we're much more resilient, and honestly much more survivable, in the event that we're ever targeted."[19] Act now to implement an ADCOP. Do not focus on what you cannot do. Focus on what you can do, and then train as you will fight.

17. **Plan the Counter-Unmanned Aerial Systems (CUAS) Fight:** Top-attack loitering munitions, unmanned combat aerial vehicles, and intelligence, surveillance, and reconnaissance UASs proliferate the modern battlespace. The enemy will use these to locate and target your CP. The enemy will attack the CP from above, so have a plan and put someone in charge of directing the CUAS fight. Use whatever CUAS weapons you have to defend your CP and coordinate this effort with other air-defense assets in the area. Have a means to report and track UASs in your battlespace. Write these methods into your SOP and practice reporting, tracking, and CUAS in training.

18. **Generate a Bias for Action:** Do not wait on someone else to solve the problem. Discuss, develop, and wargame the TTP to mask and operate a mesh CP configuration:

> Because uncertainty, degraded communications, and fleeting windows of opportunity characterize operational environments during combat, multidomain operations require disciplined initiative cultivated through a Mission Command culture. Leaders must have a bias for action and accept that some level of uncertainty is always present. Commanders who empower leaders to make rapid decisions and to accept risk within the commander's intent enable formations at echelon to adapt rapidly while maintaining unity of effort.[20]

As Gen. MacArthur warned at the beginning of this chapter, do not be too late to adapt your CP to the modern battlespace. Do not be too late in comprehending the deadly purpose of potential enemies. Do not be too late to realize the mortal danger or too late to prepare. Find a way or make one.

When it comes to Command Posts (CPs), go micro, not macro. Always mask the CP. Operating in the open is a recipe for disaster. CPs that consist of only three vehicles look more like a platoon than a CP. There will be many platoons in the battlespace. More than three vehicles become a high priority target. In this photo, soldiers set up a Tactical CP, or TAC, with three networked Strykers, for a communications exercise, on Oct. 22, 2014. Operating Mission Command under armor, masked in a forest or urban area, or on the move would be better. Train as you fight. (Photo Credit: US Army)

Human will, inspired and guided by determined leadership, is the driving force of victory in war. Technology and tactics change the methods of war, not its nature. The nature of war is a violent clash of interests between opposing wills. The best way to deter war is to be so prepared that any potential adversary will never take the risk to start one.

CHAPTER SIXTEEN

Forging Battleshock

> I am convinced that one cannot win, especially outnumbered, without attacking. But to attack on today's battlefield requires craft and cunning—a concept of offense that we've perhaps only begun to understand.[1]
>
> GENERAL DONN A. STARRY

The attack will start in 20 minutes. It is 3:20 am, August 14, 2030, and a cloudy, moonless night. Lieutenant Colonel Mike Rodriguez takes a deep breath and scans his "Mission Command Visor" (MCV) for the latest reports. With the flick of his right eye, he pulls down and reviews pertinent parts of the All Domain Common Operational Picture (ADCOP). His combined-arms battalion and supporting forces are displayed on a three-dimensional augmented-reality map, projected in front of his eyes by the visor. Critical reports from higher headquarters and his subordinate units are updated in real-time, and he sees his forces are moving as planned.

The past month has been tense, filled with sleepless nights, endless planning, combat rehearsals, and preparations. Ten days ago, his battalion moved along roads and highways through rain and lightning storms from Germany to Lithuania. The weather was horrible, with heavy rains and strong winds, but it was a blessing in disguise as they used the weather to mask their movements. Every platoon in his task force moved independently, infiltrating by multiple routes and traveling mostly at night. Dispersed movement made his units harder to observe and target, but it took time. No more than two human-operated vehicles and a dozen robotic systems were allowed in each march column. Once his teams arrived in the Close Area,[2] his tanks, armored infantry fighting vehicles, artillery, and supporting vehicles moved into the towns, cities, and surrounding forests, seeking to mask their location from enemy sensors by hiding within the noise of towns and cities, or blending into the natural camouflage of deep forests. Special radar-absorbent material camouflage covered his armored vehicles to counter the enemy's sensors. Camouflage tarps and nets reduced the infrared, thermal, and radar-band signatures of armored fighting vehicles, logistics vehicles, and command posts. His soldiers set up dozens of dummy positions as decoys in other areas to further mask the task force.

After they settled into assembly areas and camouflaged, he ordered them to "go dark," prohibiting long-range radio communications. They were only authorized to send vital messages by directed antennas with low-band millimeter-wave communications. This provided for close-range communications, and if needed, he could relay orders by passing digital messages in series, one to another in a mesh network, until everyone was included. In addition, his parent division and brigade executed a robust cyber and electronic-warfare (EW) effort to confuse the enemy, deceive their sensors, and disrupt the foe's ability to locate and target his hiding positions.

His force is the main effort of the American attack, the tip of the spear, and he was allocated multidomain assets to ensure success. His mission is to penetrate the enemy defenses and break into the enemy rear area to disrupt and destroy the enemy's artillery and anti-access/area-denial systems. The enemy comprises tank, infantry, anti-tank guided missile, and drone troops, supported by cannon and rocket artillery. His plan is to break through on a narrow front, 10 kilometers (6 miles) wide.

As he scans his zone of attack, he sees that the "Big Blue Blanket" comprising two light tactical aircraft, two helicopters, two medium-altitude, long-endurance drones, and a host of unmanned drones, all armed for the counter-unmanned aerial system (CUAS) fight, are in the air and operational. Their coverage of this "Big Blue Blanket" protects his battlespace from enemy drones and missiles. The manned aircraft will stay behind the forward line of troops, acting as mobile launching platforms for CUAS weapons, and CUAS drones will be sent forward into enemy lines as needed to defeat incoming enemy drones.

He knows, however, that the enemy is unpredictable and always has a "vote." The enemy was doing everything possible to uncover the Americans. Unable to unmask them with signal intelligence alone, the enemy had resorted to spies. Yesterday two "civilians" were arrested as they tried but failed to transmit cell-phone messages about the Americans' location. It appeared that electronic-warfare (EW) jamming of the local cell-phone network was paying off.

Enemy drones also attempted to infiltrate into his area of operations, but the American CUAS "Big Blue Blanket" shot down or disabled every attempt. At the same time, enemy sensors were actively attacked and jammed. The division and corps-level forces, specifically organized and equipped to mask the forward American units, conducted a deep operation with EW and cyber capabilities to deceive enemy sensors and disrupt the enemy's long-range precision fires (LRPF). At the same time, American sensors were focused on locating the enemy, especially his air defense, EW, command and control, and artillery units.

Rodriguez commands from inside an M1286 Mission Command Armored Multi-Purpose Vehicle (AMPV). It is truly a computer on tracks, redesigned in recent years to generate Distributed Mission Command.[3] Inside the command vehicle is a three-person crew consisting of Rodriguez, an operations officer, and a driver. Another AMPV, 400 meters behind, is the second vehicle in his command node.

This vehicle is operated by the Fire Support Officer, assisted by a senior sergeant from the operations section, a driver, and a three-soldier anti-tank guided missile and security team.

Rodriguez's mobile CP arrangement is the standard mesh nodal arrangement adopted by the US Army from lessons learned from recent wars to provide a mobile, armored, redundant, and survivable mesh CP network. The mesh arrangement consists of three CP nodes, with each node comprising two vehicles. These three pairs are always dispersed in distance but constantly connected by radio and digital communications. If any of the CP nodes is rendered ineffective, another node will take command. For this attack, Rodriguez deploys the three nodes in the form of a triangle, with two CP nodes forward and one back. His mobile CP node is to the right. Another CP node, commanded by the battalion Operations Officer (S3), Major Rob Pike, is ten kilometers (6 miles) to his left. A third CP node, commanded by the Intelligence Officer (S2), Captain Laura Adams, and a small team of staff assistants, and operating roughly in the center of the triangle formation, is eight kilometers (5 miles) behind the two leads.

Two armor companies and two infantry companies comprise his attack force. Each company has three platoons and each tank platoon consists of two manned M1A4 Abrams tanks and six robotic M7 Ripsaws acting as "loyal wingmen." These Ripsaws follow the voice or digital commands of their designated human tank commander. The M7 is latest version of the M5 Robotic Combat Vehicle (RCV).[4] The commander of each tank, therefore, has three RCV wingman capable of moving from column to line, executing a number of battle drills, and firing on designated targets. They also return fire automatically when fired upon by the enemy. The M1A4, based on the AbramsX design,[5] is a tank with enough edge-computing power[6] to sense smart systems near it and enable command of robotic systems. The tank has an unmanned turret, an automatic loader, a 120-mm cannon, and several machine guns. The crew is located inside the hull of the tank and employs see-through armor systems for superior situation awareness. Manned by a commander and a driver, the M1A4 Abrams maximizes the potential of hybrid human–robotic combat power.

Each armored infantry platoon consists of six optionally manned fighting vehicles (OMFVs) vehicles, three commanded by human crews and carrying eight infantrymen, and three completely robotic wingman systems. The robotic OMFVs, like the tanks, are commanded by a human, not controlled by an operator. Dispersed on a wide frontage, most of the vehicles are not in normal line-of-sight, but team commanders can see their location in their MCVs.

Execution of the attack was planned in a series of six pulses. First, long-range cannon artillery batteries fire a barrage of smart munitions and smoke, forcing the enemy to take cover or die. The smoke is used to prepare the battlespace for the impending drone attack. Second, initiated just after the cannons start to fire, involves a concentrated cyber and EW attack to disrupt the enemy's air defense,

EW and command-and-control (C2) systems. The third pulse is a mass attack by swarms of loitering munitions, dozens of unmanned combat aerial vehicles (UCAV), and LRPFs to destroy the air defense, EW, and C2 systems. The third pulse secures air superiority and is synchronized and executed by the AI-enabled Joint All Domain Command and Control (JADC2) kill web.

After the enemy's integrated air defenses are destroyed, the fourth pulse is activated. Networked drones enter deep into the attack zone to destroy the enemy systems. These loitering munitions are impossible to jam as they do not rely upon a controller or global-positioning satellites for navigation as they have a self-contained navigation smart system. They are also autonomous. Human control is maintained as they are networked and tracked in the ADCOP. They can be turned on, aborted, or turned off by the human commander, but otherwise they hunt within the designated "Persistent Precision Fires Box" on their own. These loitering munitions are also faster, stealthier, have better sensor packages, can loiter for up to 12 hours, and are more lethal than the drones used only a decade ago. Communicating with one another, they share target information as they hunt for targets.

The Armored Multi-Purpose Vehicle (AMPV) is one of the US Army's newest armored vehicles. The M1286 Mission Command AMPV variant accommodates a driver and commander and two workstation operators. The AMPV provides full tactical command post capabilities at brigade and battalion levels. In this photo, soldiers from the 4th Squadron, 9th US Cavalry Regiment "Dark Horse," 2nd Armored Brigade Combat Team, 1st Cavalry Division, drive through a low water crossing in an AMPV at Fort Hood, Texas, on September 24, 2022. (Maj Carson Petry, 1st CAV)

Rodriguez then initiates the fifth pulse, the close attack by his forces involving M1A4 tanks, Ripsaws, and OMFVs. With the enemy unable to come out of their defenses and mass for fear of the drones orbiting overhead, the American tanks and infantry break through and defeat the enemy. The enemy's choices are to either die in place, surrender, or run. The sixth pulse is the pursuit, using the ground force, the remaining drones, attack helicopters, and aircraft to destroy any remnants attempting to escape. Cyber and EW attacks on enemy command and computer systems continue until pulse six concludes, or the attack is called off.

Then it happened. Without warning, the enemy struck first with a terrific artillery, rocket, and missile barrage. The first strike went on for 20 minutes and amounted to a lot of noise and fire, but the masking efforts by the Americans proved effective. Enemy sensors were deceived. Incoming projectiles missed their intended targets. Many decoys were blasted to bits. False positives, identified on enemy targeting screens as real targets, but nothing but electronic shadows, were annihilated. Unfortunately, a few enemy rockets hit Americans hiding in a thick forested area and one tank, an RCV, and a manned OMFV were destroyed. Five American soldiers are killed and six wounded.

Seconds after the enemy first strike began, the American kill web launched the first pulse attack.

Inside the AMPV, Captain Bill Kent, the assistant Operations Officer (A/S3), was at Rodriguez's side, using his MCV to see the battlespace and monitor the actions of US units in real-time. Corporal Susan Kerta occupied the driver's station of the AMPV. Rodriguez, Kent, and Kerta, protected from most of the artillery fragments flying around outside, were relatively safe inside the armored vehicle.

"Pulse One now. Pulse Two initiating. Cyber and EW strike happening now," Kent said calmly. "The enemy's targeting screens should light up with more false positives. It'll look to them as if we are everywhere."

"Roger," Rodriguez acknowledged. In his MCV he witnessed some enemy UAV icons come under fire of the "Big Blue Blanket" and watched those icons dissolve. "Counter-UAS is in the fight."

A few minutes passed as far off explosions reverberated outside the AMPV.

"Pulse Three is starting. Our loitering munitions are now crossing the line of departure," Kent announced. "They are moving to designated areas to hunt inside the Fires Box."

"Affirmative," Rodriguez answered as he monitored the action with his MCV. As the American artillery blasted the enemy, Rodriguez saw the synchronized attack of a swarm of American loitering munitions and a dozen American UCAVs inside the Persistent Precision Fires Box.

Flying over the Fires Box at low altitude, these robotic systems began to search for the enemy and automatically hit hostile targets. Rodriguez knew that sending in the loitering munitions in a massed super swarm was the key to the success of the third pulse. The Fires Box was 20 kilometers (12 miles) deep and 20 kilometers wide, if all went as planned, every enemy system inside the box would be located and disabled or destroyed. The American drone attack, in conjunction with long-range precision artillery and rocket fires, would first annihilate the enemy's air defenses, disrupt the enemy's ability to command, and deny the enemy the ability to maneuver. In minutes, the enemy's air defense and EW systems were burning.

Fifteen minutes elapsed. Rodriguez's MCV indicated Pulse Two was complete. The enemy's air defense and CUAS were obliterated.

"Pulses Three and Four happening now," Kent advised.

"Affirmative. Tracking," Rodriguez replied. The tempo of the fight was picking up. It was all happening very fast. In his MCV he saw packets of loitering munitions lock onto targets and strike. LRPFs added to overall destruction as networked sensors and drones identified high-value targets for HIMARs rockets.

Twenty minutes elapsed to the rumble of distant explosions. Rodriguez waited anxiously, observing the battle shown in his MCV. The anticipation was agonizing but he knew that every minute that passed would mean fewer enemy in the battlespace. He could wait.

Through his MCV he scaled out to a larger view and surveyed the entire zone of attack. He then zeroed in on an area of interest with the flick of his eye. Multidomain information was arrayed at the far right of his view, in green to red colors depending on the status. He saw Pulse Four happening in real-time and saw the enemy defenses crumbling. Enemy icons appeared in his MCV map and then faded and dissolved as loitering munitions or missiles from UCAVs smashed into the foe.

"Pulse Four complete, sir," Kent reported.

"Acknowledged," Rodriguez said and then took a deep breath. Now was the time. He saw the reports coming in through his visor and, with the movement of his eyes, he sent a digital message to all units to begin the attack.

"Driver, move out," Rodriguez ordered over the integrated vehicle intercom system. "Stay one kilometer behind the Alpha Company on the left."

"Roger, sir," the driver, Corporal Kerta, replied.

The AMPV surged forward. Corporal Kerta turned on the thermal situational awareness driving screen and moved across the terrain as if she was moving in daylight on Main Street in her hometown. The second AMPV followed about four hundred meters behind and was in constant connection by voice and data communication with their commander.

Rodriguez received an update in his MCV that depicted the two lead companies crossing the line of departure. He smiled. The action was unfolding as they had rehearsed. His leaders knew the plan, the branch plans if something needed to change, and they understood the commander's intent if they had to act on their own without communications.

"Team Alpha has contact with the enemy," Kent reported.

"Roger," Rodriguez replied. In his MCV he saw icons representing his units in battle with the enemy and breaching the enemy defenses. A digital report from Team Alpha stated: "Enemy stunned and completely disorganized. Many surrendering. My status Green, no casualties. Continuing mission."

So far, so good, Rodriguez thought. Then, strangely, he felt a weird sensation in his gut. A loud explosion interrupted his thoughts.

There was a deafening noise. The AMPV jerked to a stop and spun to the right. The power went out, the vehicle went dark, then tilted on its side and slowly rolled over.

Secure in his seat by a seatbelt, he turned with the vehicle as objects shifted and fell around him. The turn stopped. Silence. Upside down, strapped to a seat in the dark of a flipped AMPV, Rodriguez tried to make sense of what had happened. After a moment of confusion, he regained his senses. He smelled smoke. "Crew report!"

"I'm okay," Captain Kent announced. "Damn it! Must have been a mine. No enemy mines were identified on our path, but that's probably what hit us." The captain turned on a flashlight. Smoke came from the driver's hatch. The vehicle smelled of burned propellant.

"Kerta!" Rodriguez yelled.

No answer. In a rush to check on his driver, Rodriguez struggled to unbuckle his seatbelt. When he succeeded, as the vehicle was upside down, he fell to the ceiling, banging his helmet against the metal roof. Dazed for a second, he cursed, then recovered. He unhooked his helmet connector from the on-board computer system. Crawling to the driver's compartment in the dark, he reached for Kerta, unhooked her seatbelt, and pulled her out of the driver's station. She was breathing. That was a good sign.

Kent slid to the rear hatch and struggled to get it open. He forced the handle and then pushed against the hatch with his legs. It opened. Fresh air blew into the compartment.

"Give me a hand with Kerta," Rodriguez ordered.

Kent moved toward the front of the AMPV and worked with Rodriguez to drag Kerta out of the rear hatch and onto the cool ground. In the mud, at the rear of their overturned armored vehicle, Kent looked to Kerta's wounds.

Rodriguez stood up and looked to the east. He could hear the sounds of battle, detonations, the blast of 120-mm cannon fire, the distinctive sound of the OMFV's 30-mm gun, and the rattle of machine guns. Then, to his rear, he heard a vehicle moving toward them. He turned around. It was the second AMPV in his Mission Command node, commanded by the artillery Fire Support Officer (FSO).

"Sir, is that you? Are you okay?" The FSO yelled.

"Yeah!" Rodriguez answered. "What is your status?"

"Green, sir, vehicle and comms good to go," the FSO replied. "What are your orders?"

"Get Corporal Kerta inside your track. I'll command from your vehicle. Hustle up, we're getting back into the fight."

* * *

The story above is a fictional visualization of what might one day be called Multidomain Combined-Arms Operations. It is a positive story, told with systems that are either soon to be deployed or in design, where the US military gets it right. This story has yet to be validated, but to transform idea to reality, it must start with a concept. Technology is important, and we need new systems designed around a sound warfighting concept, but technology will only take you so far.

The deadliest weapon in the world is a trained, equipped, and determined human warfighter fighting as a member of a team. Expertly trained and determined men and women, who can follow and execute Mission Command in the chaos and confusion of demolished networks, are more important than technology. Soldiers, sailors, marines, airmen, and guardians, skilled at warfighting, who can cooperate, think, act in time, and will never give up, will always be our asymmetric advantage. No technology can replace teams of well-trained, prepared, and determined warfighters, but such teams with the best technology and tactics can generate overmatch at the decisive point. Properly applied, this combination generates battleshock. To do this we must **lead**, **design**, **train**, **fight**, and **support** to **win**.

Lead

Command in war is always difficult. Thus, we admire great military leaders and, to this day, study the leadership and actions of Alexander the Great, Julius Caesar, and Napoleon. Military genius was able then, and now, to peer through the fog of war, solve immediate problems, and win victories. War is more complex today than in Alexander's time, but leadership in war remains remarkably similar. Leaders must earn the trust of those they lead, train their forces, make decisions, act in time, and lead by example. In war, leaders must operate without complete information, in what Clausewitz termed the "fog of war," and, with imperfect information, coordinate the military means at their disposal in time, space, and purpose.

Since human warfighters are central to winning at war, the military culture around which soldiers, sailors, airmen, marines, and guardians are incorporated matters. Top-down, orders-intensive leadership cultures such as those of Russia, China, Iran, and North Korea require human warfighters to obey orders and execute instructions according to plan. These military cultures put a premium on maximizing the plan and their efforts run astray when the plan is rendered irrelevant by circumstances, something that always happens in war.

Combined arms wins, but combined-arms tactics must change to be successful in today's transparent, lethal, and precision battlespace. US Army soldiers with the 2nd Battalion, 3rd Field Artillery Regiment, 1st Armored Brigade Combat Team, 1st Armored Division, shoot an M109 Paladin as part of a battalion-level artillery qualification exercise at the Dona Ana Range Complex, New Mexico, on March 8, 2021. (US Army photo by Pfc Luis Santiago)

A good example of this is the Russian river-crossing operation of Siverskyi Donets River in the Donbas region on May 10, 2022. In order to advance and outflank the Ukrainian defenders, the Russians conducted an assault river crossing. Using multiple sensors, including human scouts, drones and satellite imagery, the Ukrainians uncovered what the Russians were up to and counterattacked with a fire strike of artillery and drones to defeat the crossing. Determined, and sticking to the plan, the Russians tried five times to cross the river, taking heavy casualties every time. Russian commanders at the point of decision were not authorized to alter the plan and had to attack across the river even after each attempt appeared to be nothing more than suicide. You might admire their determination, but also question their judgement as this became one of the costliest operations in the war to this point. The Russian top-down command culture created the disaster at the Siverksyi crossing, losing more than 80 vehicles and hundreds of soldiers. The Russians, Chinese, Iranians, and North Koreans do not believe that "who thinks, wins." They believe that "who obeys, wins," and they extol episodes in their military history where this stoic determination is exhibited. This approach is predictable and scientific as subordinate units are expected to do exactly as ordered.

Western military culture generally embraces a different approach which is explained by the concept "Mission Command." Mission Command contends that human decision making at the point of contact generates a flexible and winning approach. In Mission Command, leaders are expected to think and act in accordance with their commander's intent to make decisions in real-time to win the fight. In the chaos and fog of war, when leaders are killed, Mission Command-enabled units will continue to operate and succeed. Mission Command embraces the idea "who thinks, wins." Operating with a shared vision of the battlespace, by understanding the intent of the commanders two echelons higher, leaders and units that are excellent at Mission Command gain a time and decision advantage over their "I only follow orders" opponents. Leaders embracing Mission Command must understand what it is, be able to articulate it clearly, teach it, and live it. A leader must define Mission Command in the leader's own words, visualize it in the leader's mind's eye, and live by it. Understanding Mission Command, practicing it in garrison life, and reinforcing it in training, is vital to using Mission Command in combat.

Technology can enhance how leaders interact. In ancient times, leaders required face-to-face interaction, or used messengers with verbal or written orders. As advances in communication developed, speedier means were used such as flag signals, the telegraph, wired telephones, radio, and now digital communication by a 5G internet cloud. Today's digital advances allow commanders to be placeless in the battlespace. Virtual reality and augmented reality (AR) can be used as powerful tools of Mission Command. As mentioned at the beginning of the chapter, no machine can replace a trained, equipped, and determined human warfighter. The best efforts in the near term are to enable humans by technology and create a human–robotic hybrid force. Mission Command, the approach to command and control that empowers subordinate decision making and decentralized execution appropriate to the situation can be enhanced with technology. A centaur approach offers the best solution, with human and technology working seamlessly together. In a Centaur Mission Command set-up, the artificial intelligence (AI) embedded in edge computing and visualized through a HoloLens-like AR visor would enable the *coup d'oeil* commanders need to operate in the multidomain battlespace. Imagine a helmet or visor, embracing a Centaur Mission Command concept, where commanders at all echelons are connected to see the battlespace in real-time, across multiple domains, at the flick of an eye or the touch of a control. As J. F. C. Fuller once said, "The more mechanical became the weapons with which we fight, the less mechanical must be the spirit which controls them."[7] To move forward and apply technologies in time, leaders must understand the impact of new technologies in war, in order to avoid surprise and defeat. Understanding the nature of war, and its changing methods, is therefore a critical leadership task.

If we change our views about command posts (CP) to organize them around Distributed Mission Command, then we can deliver the function of a CP as a service, not a place. Imagine if we established CPs in a mesh network of command nodes, each comprising 2–4 networked armored vehicles, that allowed any commander, from battalion to corps, to operate seamlessly from any of the nodes in the mesh network? The commander would traverse from node to node to enhance command presence and leadership. If one node is disrupted, another takes over, and the new acting commander commands the force. This requires resilient, dynamic, secure communications, and a shared All Domain Common Operational Picture (ADCOP).

Design

As the methods of war change, the design and development of weapons must change. New circumstances require new weapons. Today, it is difficult to mask in the battlespace. Enemy optical, thermal, electronic, or acoustic sensors will find you. Tomorrow, it may be impossible when quantum sensing is added to the list. Masking, therefore, must become a critical design parameter for all future military systems. Flooding the battlespace with many "false" targets, using cyber operations or physical decoys, is another way to mask.

Loitering munitions provide the means to swarm the enemy and will become the weapon of choice in tomorrow's conflicts. They provide an inexpensive, disposable, and effective means to provide precision striking power without human loss for the attacker. For these reasons they are sent where other forces are not, as they are modern-day kamikazes. Loitering munitions also offer a sense-and-strike ability that can work with or without other sensor networks to provide intelligence and surveillance of the battlespace. Networked loitering munitions, used en masse and integrated into a super swarm assault, will become tomorrow's breakthrough method. Faster, stealthier, more precise, and more powerful loitering munitions are arriving in the battlespace and will become the weapon of choice.

For every weapon, we must design and develop counter-weapons. As a result of the success of loitering munitions and UCAVs, CUAS design and development is now as urgent and as important as it was to find a means to protect the US Navy's fleet against kamikazes in the last years of World War II. Creating a continuous air-defense coverage over friendly forces, the "Big Blue Blanket" concept, worked in World War II and provides an outline for today's solution. Adopting a high-low mix, that takes advantage of the mobility and firepower of light tactical aircraft and uses these aircraft as communication and control nodes in an integrated mesh air-defense network for all CUAS systems in range, could provide the centerpiece of today's version of the "Big Blue Blanket."

To fight modern wars, we need a host of robotic systems commanded, not controlled, by human warfighters. Consider a Tank Robot Unified Servant Team (TRUST) concept, where the human commands the robot—human-on-the-loop—to execute a playbook of several simple commands to robot servant (RS) fighting armored vehicles such as "move in column," "move into firing line," and "engage targets." This is dramatically different than controlling a robotic vehicle. The human-on-the-loop does not control the robot. The RS is a servant to the manned tank. Commands to move the robot to exact locations would be the exception. Within the limitations of AI, the human would merely designate a position for the robot, much as a platoon leader issues commands to human warfighters. The robot would then pick the best path and arrive at the designated destination using on-board AI.

Fighting against peer adversaries requires smarter vehicles and weapons designed around edge-computing power.[8] We urgently need this for new ground vehicles. Mission success in large-scale ground combat operations requires integrated and synchronized multidomain (air, land, maritime, cyber, and space) operations in near real-time. Joint military forces urgently need a shared conceptual framework for commanders to visualize an ADCOP. This could manifest in a commander's "dashboard" to allow leaders to "see themselves, the enemy, and other domains" in the battlespace. A common operational picture is a single identical display, of a relevant (operational) information (e.g., position of friendly forces, enemy forces, and positions and status of key infrastructure items such as bridges, roads, etc.) shared by more than one command. An ADCOP would facilitate collaborative planning and combined execution and assist all echelons to achieve situational awareness by sharing this picture across the joint, all-domain force.[9] Transforming intelligent systems into a mesh network of nodes will enable Distributed Mission Command and improve maneuver and striking power in the transparent battlespace.

Train

Mission Command, however, is a two-edged sword that can cut both ways. It requires trained warfighters and leaders who know how to conduct operations using Mission Command. The level of leadership, experience, and trust to lead with Mission Command takes time to develop and attain. As casualties occur in combat, the ability to continue to execute Mission Command is as difficult as a sports team that experiences the replacement of key players with rookies. Failure can occur when untrained leaders and units execute operations as they think is best, when in fact, their actions are not in coherence with their higher commander's intent. Training must mirror this to include the opportunity for junior leaders to take command at critical moments by removing the commander. The number of training iterations—in constructive, virtual, and live simulation—must reach a level of practice that Mission Command becomes second nature. Most militaries find Mission Command too difficult to develop and employ and therefore default to an "orders-intensive" approach to command.

The purpose of training is to develop individuals, teams, and units that can win in combat. New technologies can enhance training for Mission Command.

The AeroVironment Switchblade 600 shown here is a battle-tested loitering munition that can fly out to 40 kilometers (25 miles) in 20 minutes, then loiter for another 20 minutes seeking targets (giving it an 80-kilometer [50-mile] total range). It attacks at 185 km/h (115 mph) with a warhead similar to a Javelin anti-tank guided missile. Networking Switchblades into a super swarm is the next step. (AeroVironment)

Simulations can be used to create a battle school where Mission Command is frequently tested in realistic conditions. For example, in a typical two-year training cycle, battalions in the US Army normally engage in several battalion-level constructive and virtual battle simulations but only one "live" training rotation to a combat training center (CTC). Live training is expensive and will always have a lower frequency than constructive and virtual training. Mission Command, therefore, must be built with rigorous constructive and virtual experiences, with a frequency that builds mastery. This requires focus on warfighting, something that most armies find difficult until the bugle sounds for battle. By then, it is too late; experience has historically been gained through combat. If we expect to enter combat ready on Day 1, we must change this paradigm.

Let us imagine the training regime for a US Army Combined Arms Battalion that is scheduled to conduct a training rotation at the National Training Center in 14 months. The battalion commander has much to prepare for. His unit, which is constantly changing leaders due to the US Army's individual replacement policy, must train at a frantic frequency to keep leaders and units at the level to execute Mission Command. Many constructive simulations are available to support this effort, from paper combat-decision games to sophisticated computer simulations such as VBS3 and VBS4. Constructive simulations primarily train the mind. Virtual simulation is conducted by sophisticated training setups that train the mind and body, such as aircraft, tank, vehicle, and squad simulators. Conducting constructive multi-echelon training simulations, once a week, with a smattering of virtual exercises, would build a repertoire of a maximum of roughly 64 iterations. Reduce this by half to complete other requirements and now you are down to 32 iterations. In sports, a winning team does the reps needed to master its moves. How many iterations will it take to gain mastery at collective warfighting skills? How many times must a leader plan, prepare, and execute various missions to gain excellence? Every commander must ask this question and plan accordingly. The US military is the most simulation-rich military force in the world. The problem is it does not maximize the existing simulation capability. We will not rise to our expectations. We will fall to the level of our training. We have many excuses for not training like we should. We aren't doing the reps and this is why it will be difficult to execute Mission Command as we visualize we should when the shooting starts.

Fight

Today, we see the impact of deployed and emerging technology on the changing methods of war. Leaders adapt to the changing methods and use them to their advantage. In the past several years, the Second Nagorno-Karabakh War (2020), the Israel–Hamas War (2021) and the ongoing Russia–Ukraine War (2022–) have

US Marines launch a small unmanned aerial system at a mortar range on Camp Lejeune, North Carolina, January 17, 2023. The integration of unmanned aircraft systems into infantry units will help the Marine Corps be more lethal and effective on a small-unit level. (US Marine Corps photo by Cpl Michael Virtue)

demonstrated the use of robotic systems, primarily UCAVs and loitering munitions, and the devastating effect of the combination of sensors and long-range precision fires. These facts require leaders to address the topics laid out in this book and prepare for war in a transparent battlespace, where the enemy will most likely have the first-strike advantage, and where rapid technological change will require thinking leaders more than ever in the history of warfare.

Leaders preparing warfighters for the next war must adapt their efforts to these new realities. The first task is to read, dialogue, discuss, and learn how the changing methods of war will affect the unit you lead. The next is to do something about it. Train in a transparent battlespace where the target that sticks out gets hammered. Learn to embrace masking, the full spectrum, multidomain effort to **deceive** enemy sensors and **disrupt** enemy targeting, in every training event. Plan and train as if the enemy will have the first-strike advantage, and take measures to maximize masking, protection, dispersion, and early warning to minimize this enemy strategy. If an electronic military system communicates, it can be jammed. If it has an electronic signature, it will be identified and located by sensors. This works both ways, for us and the enemy. We must reduce our signatures by every means possible. If you command forces on the island of Guam, for instance, and know that everything you command can be seen by Chinese multidomain sensors and are in range of Chinese missiles, what are you doing to prepare for the possibility of a surprise attack? The dynamics of conflict are too complex to be reduced to a formula, hence there is no straightforward equation or set of instructions that can prepare a command for combat. If change is the only constant in both leadership and war, then leaders must understand the changing methods of war and prepare in time, or risk being victims of the tragic lessons of recent conflicts.

Support

We must always train as we expect to fight. US forces must be able to support forces in a contested battlespace, where there are no sanctuaries. Half-measures will get us killed. In the Russia–Ukraine War, the failures of Russian logistics have been demonstrated to the world. Rotting tires, old rations, lack of fuel, inoperable radios, and a hundred other shortcomings due to inept leadership and corruption, have plagued Russian logistical support of their combat forces. During recent fighting, Ukrainian long-range precision fires have targeted Russian ammunition dumps and fuel and supply depots with spectacular results. Russian logistics units were not ready for war and the Russian Army paid for these flaws in killed and wounded. We must question if our logistics units are ready for today's lethal battlespace? In 2022, I visited the US Army's National Training Center to observe a training rotation and, specifically, investigate the survivability of the training unit's CP. The logistics units supporting this rotation were severely challenged. It was clear to me, and everyone

A US soldier, assigned to the 2nd Cavalry Regiment, engages the enemy during *Dragoon Ready 23* at the Joint Multinational Readiness Center in Hohenfels, Germany, February 1, 2023. The exercise trains the regiment in its mission-essential tasks in support of unified land operations to enhance proficiency and improve interoperability with NATO allies. (US Army Photo by Staff Sgt Jose H. Rodriguez)

who observed the training, that the US Army must spend more energy and effort preparing its support units for war.

Supporters are also fighters. Maintenance and supply personnel must train under the conditions of the warfighting disrupters listed above. There are no sanctuaries in the battlespace. Every position is observed and in range of long-range precision fires. As mentioned earlier, the target that sticks out, gets hammered. Support-unit CP, ammunition and fuel dumps, and maintenance facilities offer the enemy lucrative targets that are easy to find and difficult to defend. Masking these vital warfighting enablers is paramount. If the enemy destroys support-unit command and control, turns ammunition and fuel dumps into raging fires, and pummels maintenance efforts with drones and top-attack munitions, it will not matter how powerful the fighting units are. Without fuel, ammunition, and supplies, any military force is turned powerless. We must train our support forces to meet this challenge. Medical-evacuation capabilities must extend the "golden hour" into a "golden day" as we will no longer have the permissive environment to evacuate our wounded as we have in the counterinsurgency wars of the past decades.

In short, support units need experienced leadership and a higher level of training than they demonstrate today. They will require extensive training iterations in constructive, virtual, and live simulation to meet the demands of modern combat. We will no longer be able to concentrate our support teams in well-defended forward operating bases. In the modern battlespace, support and logistic units should be able to operate dispersed and every effort should be made to increase their mobility and masking. We need to think differently and determine how to spread out support units in a mesh-network organization and increase their survivability by protecting them with air-defense and anti-projectile systems such as the Israeli-made Iron Dome and counter-rocket artillery and mortar systems. Establishing a "Big Blue Blanket" over support areas is vital. We must also practice hardening positions and digging in. Support units need a complete revitalization of tactics, techniques, and procedures to survive and support the warfighting elements. These are critical issues that demand immediate action.

Win

To win the next war, we must study and act on the top warfighting disrupters outlined in this book and, after a thorough evaluation, adapt that knowledge to our execution of multidomain operations (MDO) and combined-arms operations. The way the US Army conducts combined-arms operations today must adapt to the changing methods of war. How we plan, prepare, and execute combined arms must change. Combined arms is not a new concept. Modern armies recruit, equip, and train their ground forces to conduct combined-arms operations. Ancient armies did the same. The orchestration of various tactical capabilities, from the use of spearmen with javelin throwers and cavalry in the times of the Roman legions, to the application of cavalry, infantry, and artillery in the wars of Napoleon, are examples of combined arms. Each singular force had a distinct purpose. Infantry held ground and assaulted to take ground. Artillery pummeled the enemy into submission and fought a battle against the enemy's artillery. Cavalry screened the front and conducted reconnaissance, sought to defeat the enemy's cavalry, and conducted fast-flowing maneuvers and pursuits.

Combined-arms warfare is the simultaneous application of combat, combat support, and combat-service support toward a common goal. Combined-arms warfare produces effects greater than the sum of the individual parts. The combined-arms team strives to conduct fully integrated operations in the dimensions of time, space, purpose, and resources to confuse, demoralize, and destroy the enemy with the coordinated impact of combat power. Today, the means of war have expanded a thousandfold since the days of Napoleon, and so have the ways to apply combined arms.

Multidomain combined arms is the simultaneous convergence of the effects from land, sea, air, cyber, and space to facilitate combined-arms operations. Multidomain combined-arms maneuver directs the convergence of the joint force to overwhelm enemy forces inside the designated MDO Breakthrough Zone to create battleshock. Coordinating and synchronizing these efforts is extremely complex and will require Distributed Mission Command and expert systems such as are being developed with JADC2. Dispersing, attacking, and massing will require that we become excellent at masking. Generating MDO combined arms will take hard training and skilled leadership that is excellent at Distributed Mission Command.

Very few armies in the world will be capable of MDO combined arms. Those that do will hold a tremendous advantage, similar to the *blitzkrieg* of 1940. The friction of war, what Clausewitz defined as "the random and unpredictable events within a given conflict that cannot be foreseen," applies today and will wear military forces down over time. The default position in warfare is positional warfare, static defense, and bloody attrition. Without maneuver, war drags on and on, producing heavy casualties on both sides with no decision in sight. This occurred in World War I, when firepower technology raced ahead of maneuver technology and tactical thinking. Today we are in a similar situation with technology and tactics as in the Great War. When friction increases, war reverts to positional fighting as we see in the war in Ukraine. Digging in can save you, since artillery and precision-fire weapons proliferate the battlespace. Digging in can make a position transform from weak to strong, but it cannot win the war. Only offensive action can do that. Once in battle, you must attack, gain the initiative, and impose your will on the enemy to bring about a decisive conclusion.

The Next War

New means of war stimulate new ways of thinking. The wars of recent years have shown us the changing face of conflict. The observations and lessons from these are a clarion call to action. They are our wake-up call. From my study of these wars, I developed the nine disrupters outlined in this book. It is clear from these wars that we must avoid protracted conflicts. We must learn from our long, painful experience in Iraq and Afghanistan. If we do, we should focus on winning wars rapidly and decisively, as we did in *Desert Storm* in 1991. War today, however, is very different from then and what we did in *Desert Storm* cannot be replicated. The enemy will not allow us to build up an overwhelming force in a sanctuary and await our attack. As described in these chapters, the enemy will have a first-strike advantage and we must change our training and thinking to meet that challenge.

This is a time of trouble with pandemics, economic problems, and war. The "when and where" of America's next war is an exercise in prognostication, but there are clear warning signs we should observe. Today, the US is in a hybrid war with Russia, as America and other nations arm Ukraine to defend against Russian aggression. Tensions between the US and NATO versus Russia are growing as Russia finds it harder to win in Ukraine. Without notice, the fighting in Ukraine could escalate into a shooting war between Russia and NATO, either by mistake or design.

On the other side of the globe, China is a growing threat to the US and its interests. The risks of war in the Pacific are high. "China sees itself in a state of war with its adversaries, but its adversaries do not."[10] China is threatening Taiwan and promising to take the country by any means, including military invasion. In November 2022, China witnessed the largest internal demonstrations since the 1989 revolt in Tiananmen Square, the "White Paper Protest." The Chinese Communist Party silently arrested or eliminated every leader and many others identified in these demonstrations, proving its surveillance state is beyond even George Orwell's dystopian novel *1984*.[11] In spite of unrest at home, China is modernizing its military forces to take Taiwan. "This is not a matter of if they will invade, it's a matter of when they will invade."[12] An attempt to conquer Taiwan would trigger a US and allied response, pitting the US against a peer power. If so, North Korea might also get involved. US forces in the Pacific and the Republic of Korea must always be on guard as China or North Korea can hit them with a first strike with little warning.

Leadership Matters. The methods of war are ever-changing. Read to lead. Who thinks, wins! A US Army infantry squad leader and machine gunner engage enemy targets during a combined-arms live-fire exercise at Bemowo Piskie, Poland, on February 27, 2023. (US Army National Guard photo by Sgt Lianne M. Hirano)

In other areas, such as Nagorno-Karabakh, a new war could break out anytime between Azerbaijan and Armenia. Azerbaijan and Turkey are allies, as Armenia is allied with Russia and Iran, so a regional war in the Caucasus looms. Add to this Russia's growing instability, due to its demonstrated weaknesses in the Russia–Ukraine War. New wars could erupt in any of the old Soviet republics. Checked from internecine and ethnic fighting by Moscow's influence, these independent countries may choose war or dissolve into civil wars to address old grievances as Russia loses its grip. If Russia fails in Ukraine, the Russian Federation could collapse. The fall of empires has caused wars in the past. The next few years will be dangerous and filled with potential wars and rumors of war. Now is the time to raise awareness, sharpen foresight, and prepare. We cannot wait for 2030 or 2035 to be ready.

We must reimagine how we fight. Technology is vital and humans are more important than hardware, but only if those humans think and prepare. We will not rise to the level of our most exquisite technology; we will fall to the level of our collective training, using what we have when the war starts. If we do not learn from recent wars, we will not get a second chance.

The future will be here sooner than you think. All these potential conflicts are accelerated by new technologies. We seem to be at the precipice of a major war. If we prepare now, we may be able to convince potential adversaries to think twice before launching a war. Deterring war requires the enemy to believe he cannot win a conflict and should not even attempt to attack us or our allies. If deterrence fails, we must win as speedily as possible, as long wars are the ruin of nations. To win wars against peer enemies we need a robust dialogue about the nine disrupters described in this book. The clocks are "striking thirteen" and time is running out to make a difference. We need to think and act differently. If we do not study each of these disrupters, and learn the lessons of recent wars, we will pay a heavy price in blood in the next war. If, on the other hand, we work hard to gain foresight, and act in time to develop innovative solutions that include new ideas, tactics, and equipment, we can deter wars and successfully defend our allies and our nation. Leadership is the author of victory and the publisher of peace. It is up to us, as leaders, to think, act, and be ready for the next war.

Endnotes

1. Ulysses S. Grant, as quoted in James F. Rusling, *Men and Things I Saw in Civil War Days* (New York: Eaton and Mains, 1899), 137. Note: Colonel James F. Rusling of the quartermaster general's staff recalled that in the winter of 1863–64, a quartermaster officer approached Grant for approval of millions of dollars of expenditures for the coming Atlanta campaign, and Grant approved the expenditure after briefly examining the papers involved. Questioning Grant's swift decision, the officer asked him if he was sure he was right. Grant replied, "No, I am not, but in war, anything is better than indecision. We must decide. If I am wrong, we shall soon find it out and can do the other thing. But not to decide wastes both time and money and may ruin everything."

Preface

1. Bertrand Russell, *The A B C of Relativity* (New York: Harper & Brothers, 1925), 167.
2. The idiom "Who thinks, wins" is inspired by the British Secret Air Service motto, "Who dares, wins" and from Bryce G. Hoffman's excellent book, *Red Teaming: How Your Business Can Conquer the Competition by Challenging Everything* (New York: Crown Business, 2017), 7.
3. The story of the Trojan Horse is told in Homer's *Odyssey* and in Virgil's *Aeneid*. In the *Aeneid*, Odysseus is called by the Roman name "Ulysses." See: M. Clarke (1970), *The Aeneid of Virgil*, translated by John Dryden. Edited with introduction and notes by Robert Fitzgerald. *The Aeneid by Virgil* (London: Collier-Macmillan, 1968), accessed at http://faculty.sgc.edu/rkelley/The%20Aeneid.pdf.
4. Patrick Sanders, "Chief of the General Staff Speech at RUSI Land Warfare Conference," *Gov.UK*, June 28, 2022, at https://www.gov.uk/government/speeches/chief-the-general-staff-speech-at-rusi-land-warfare-conference.
5. Note: As the head of the US Army Training and Doctrine Command (TRADOC), Gen. DePuty studied the Yom Kippur war and revised US Army doctrine. As a result of this assessment, a new doctrine based, called the Active Defense, was published in FM 100-5 Operations in 1976. Gen. Starry, who succeeded Gen. DePuy as the commander of TRADOC, changed doctrine again by emphasizing the joint employment of air and land forces and emphasising maneuver rather than attrition. The result of Starry's efforts was AirLand Battle, an upgraded version of FM100-5 that was released in 1982. AirLand Battle was the overall conceptual framework that formed the basis of the US Army's warfighting doctrine from 1982 into the early 2000s and demonstrated successfully during Operation *Desert Storm* in 1991.
6. See the Glossary for a detailed definition of Artificial Intelligence (AI) in a 2018 US Congressional report.
7. Todd Neikirk, "The Germans Had High Hopes Goliath for the Goliath Tracked Mine," *War History Online*, accessed June 6, 2022, at https://www.warhistoryonline.com/world-war-ii/goliath-tracked-mine.html?chrome=1. Note: the German army deployed both an electric and gas engine version of the Goliath.

8. Jen Judson, "US Army Adopts New Multi-domain Operations Doctrine," *Defense News*, accessed October 22, 2022, at https://www.defensenews.com/land/2022/10/10/us-army-adopts-new-multidomain-operations-doctrine/.
9. Joint Air Power Competence Center (JAPCC), "All Domain Operations in a Combined Environment," North Atlantic Treaty Organization JAPCC, September 2021, at https://www.japcc.org/flyers/all-domain-operations-in-a-combined-environment/.
10. US Congress, "Testimony of Peter Singer to the House Armed Services Committee Subcommittee on Cyber, Information Technologies, and Innovation The Future of War: Is the Pentagon Prepared to Deter and Defeat America's Adversaries," Washington, D.C.: US House of Representatives, Armed Services Committee, February 9, 2023.

Chapter 1

1. Roberta Morgan Wohlstetter, *Pearl Harbor, Warning and Decision* (Stanford: Stanford University Press; 1962), 387.
2. George Sylvester Viereck, "What Life Means to Einstein: An Interview by George Sylvester Viereck," Indianapolis: *Saturday Evening Post* Society, October 26, 1929, 17.
3. Command History. US Pacific Fleet website, US Navy, at https://www.cpf.navy.mil/About-Us/Command-History/; and Bob Bryant, "Aircraft Carrier Operations During WW2," *WWII Database*, at https://ww2db.com/other.php?other_id=49.
4. A shock drone is a term sometimes used in Russian military articles to describe a UAV or loitering munition that is designed for both reconnaissance and strike missions.
5. Those who wish to study this war should read my book *7 Seconds to Die: A Military Analysis of the Second Nagorno-Karabakh War and the Future of Warfighting*. See https://www.casematepublishers.com/7-seconds-to-die.html.
6. Dontavian Harrison, "AUSA 2021: Secretary of the Army's Keynote Speech 11 October 2021." *army.mil*, accessed October 14, 2021, at https://www.army.mil/article/251180/ausa_2021_secretary_of_the_armys_keynote_speech_11_october_2021.
7. Anna Ahronheim, "Israel's Operation Against Hamas was the World's First AI War," *The Jerusalem Post*, May 26, 2021.
8. Harun Yilmaz, "No, Russia will not invade Ukraine: A large-scale military operation does not fit into Moscow's cost-benefit calculus," *Al Jazeera*, February 9, 2022, at https://www.aljazeera.com/opinions/2022/2/9/no-russia-will-not-invade-ukraine; and Frank Gardner, "Ukraine crisis: Five reasons why Putin Might Not Invade," *BBC News*, February 21, 2022, athttps://www.bbc.com/news/world-europe-60468264.
9. Jacqui Heinrich, Adam Sabes, "Gen. Milley says Kyiv could fall within 72 hours if Russia Decides to Invade Ukraine," *Fox News*, February 5, 2022, at https://www.foxnews.com/us/gen-milley-says-kyiv-could-fall-within-72-hours-if-russia-decides-to-invade-ukraine-sources.
10. Geoff Ball, Chad Skaggs, and USMC authors et al, "Signature Management (SIGMAN) Camouflage SOP: A Guide to Reduce Physical Signature Under UAS," Twenty-Nine Palms: The Warfighting Society, August 2020, at http://www.2ndbn5thmar.com/camouflage/SIGMAN%20Camouflage%20SOP.pdf, 1.
11. Mission Command is the Army's approach to command and control that empowers subordinate decision making and decentralized execution appropriate to the situation. See: Army Doctrine Publication No. 6-0, *Mission Command: Command and Control of Army Forces*, Washington, D.C: Headquarters Department of the Army, July 31, 2019, 1–3.

12. US Army, "The Operational Environment and the Changing Character of Warfare," Fort Eustis: TRADOC Pam 525-92, October 2019, 24.

Chapter 2

1. Mark A. Milley, "AUSA Eisenhower Luncheon, 4 October 2016" (speech, Association of the United States Army [AUSA]), Washington, D.C., October 4, 2016.
2. Raphael Satter and Dmytro Vlasov, "Ukraine Soldiers Bombarded by 'Pinpoint Propaganda' Texts," *AP News*, May 11, 2017, at https://apnews.com/article/technology-europe-ukraine-only-on-ap-9a564a5f64e847d1a50938035ea64b8f.
3. David Axe, "The Ukrainian Army Learned The Hard Way—Don't Idle Your Tanks When The Russians Are Nearby," *Forbes Magazine*, August 5, 2020, at https://www.forbes.com/sites/davidaxe/2020/08/05/the-ukrainian-army-learned-the-hard-way-dont-idle-your-tanks-when-the-russians-are-nearby/.
4. Phillip Karber, "Lessons Learned From the Russo-Ukraine War," Historical Lessons Learned Workshop by Johns Hopkins Applied Physics Laboratory and the US Army Capabilities Center (ARCIC), July 6, 2015, at https://www.researchgate.net/publication/316122469_Karber_RUS-UKR_War_Lessons_Learned.
5. Yury Butusov, "The Russian Army killed 37 Ukrainian Soldiers near Zelenopillia," *Censor.NET*, Ukrainian news portal, July 11, 2014, at https://censor.net/ru/r343516.
6. Nomophobia is an abbreviated form of "no-mobile-phone phobia." The term was first coined in a 2008 study that was commissioned by the UK Postal Office. See Kendra Cherry, "Nomophobia: The Fear of Being Without Your Phone," *Verywell Mind*, February 25, 2020, at https://www.verywellmind.com/nomophobia-the-fear-of-being-without-your-phone-4781725.
7. Liam Collins, "Russia gives Lessons in Electronic Warfare," *Association of the US Army*, July 26, 2018, at https://www.ausa.org/articles/russia-gives-lessons-electronic-warfare.
8. Lt. Gen. Sergei Sevryukov, "Statement by First Deputy Chief of Main Military-Political Directorate of Russian Armed Forces Lieutenant General Sergei Sevryukov," *mil.ru*. Russian Ministry of Defense Video, January 4, 2022.
9. Malcolm Parkinson, "The Artist at War," *Prologue Magazine* in The National Archives, Vol. 44, No. 1, Spring 2012, at https://www.archives.gov/publications/prologue/2012/spring/camouflage.html.
10. Linda Rodriguez McRobbie, "When the British Wanted to Camouflage Their Warships, They Made Them Dazzle," *Smithsonian Magazine*, April 7, 2016.
11. Hugh. B. Cott, *Adaptive Coloration in Animals*, Methuen, Oxford University Press, 1940. Hugh Cott established these methods of camouflage from nature: 1. merging, e.g., hare, polar bear; 2. disruption, e.g., ringed plover; 3. disguise, e.g., stick insect; 4. mis-direction, e.g., butterfly and fish eyespots; 5. dazzle, e.g., some grasshoppers; 6. decoy, e.g., angler fish; 7. smokescreen, e.g., cuttlefish; 8. the dummy, e.g., flies, ants; 9. false display of strength, e.g., toads, lizards, at https://www.liquisearch.com/hugh_b_cott/camouflage_research.
12. Dan Parsons, "What Ukraine Is Teaching US Army Generals About Future Combat: US soldiers can expect to be under constant enemy surveillance and threatened by long-range precision artillery in the next war," *The War Zone*, October 14, 2022, at https://www.thedrive.com/the-war-zone/what-ukraine-is-teaching-u-s-army-generals-about-future-combat.
13. Mykhailo Podolyak, "If something is launched into other countries' airspace, sooner or later unknown flying objects will return to (their) departure point." Twitter, @Podolyak_M, December 5, 2022.

14. John Johnson, "Analysis of Image Forming Systems," Ft. Belvoir: Image Intensifier Symposium, AD 220160, Warfare Electrical Engineering Department, US Army Research and Development Laboratories, October 6–7, 1958, 249–73, at https://home.cis.rit.edu/~cnspci/references/johnson1958.pdf.
15. Dmitry Litovkin, "Infrared and Invisibility: Russia's New Tanks Top Up on Technology," *Russia Beyond*, May 17, 2017, at https://www.rbth.com/defence/2017/05/17/infrared-and-invisibility-russias-new-tanks-top-up-on-technology_764601.
16. Jennifer Chu, "MIT engineers configure RFID tags to work as sensors, Platform may enable continuous, low-cost, reliable devices that detect chemicals in the environment," *MIT News*, Massachusetts Institute of Technology, June 14, 2018; and LaPorta, James, et al. "Military Units Track Guns Using Tech that Could Aid Foes," *The Associated Press*, September 30, 2021.
17. Swati Khandelwal, "How A Drone Can Infiltrate Your Network by Hovering Outside the Building," *The Hacker News*, October 7, 2015, at https://thehackernews.com/2015/10/hack-drones-computer.html.
18. Sam Cranny-Evans and Dr. Thomas Withington, "Russian Comms in Ukraine: A World of Hertz," *Russia Report*, March 9, 2022, at https://rusi.org/explore-our-research/publications/commentary/russian-comms-ukraine-world-hertz.
19. Stephen Chen, "Chinese Team Says Quantum Physics Project Moves Radar Closer to Detecting Stealth Aircraft," *South China Morning Post*, September 3, 2021, at https://www.scmp.com/news/china/science/article/3147309/chinese-team-says-quantum-physics-project-moves-radar-closer.
20. US Army doctrine establishes nine principles of war: Objective, Offensive, Mass, Economy of Force, Maneuver, Unity of Command, Security, Surprise, and Simplicity. Joint doctrine adds three principles of operations (perseverance, legitimacy, and restraint) to the traditional nine principles of war to account for operations other than conventional large-scale combat, such as peacekeeping and counterinsurgency. See: US Army, *Field Manual 3-0, Operations* (Washington, D.C.: Headquarters Department of the Army, 2022). Appendix A.
21. Kateryna Stepanenko, et al. "Russian Offensive Campaign Assessment, September 30," Institute for the Study of War, September 30, 2022, at https://www.iswresearch.org/2022/09/.

Chapter 3

1. Sun Tzu, *The Art of War*, (Lionel Giles Trans.), Project Gutenberg, (Original work published London: Luzac and Company, 1910), 2004 by project Gutenberg, at https://ia600502.us.archive.org/12/items/TheArtOfWarBySunTzu/ArtOfWar.pdf, 65.
2. Lester W. Grau and Charles K. Bartles, "The Russian Reconnaissance Fire Complex Comes of Age," Oxford: The Changing Character of War Centre Pembroke College, University of Oxford, With Axel and Margaret Ax:son Johnson Foundation, May 2018, at https://static1.squarespace.com/static/55faab67e4b0914105347194/t/5b17fd67562fa70b3ae0dd24/1528298869210/The+Russian+Reconnaissance+Fire+Complex+Comes+of+Age.pdf.
3. Mark F. Cancian, Matthew Cancian and Eric Heginbotham, *The First Battle of the Next War: Wargaming a Chinese Invasion of Taiwan*, Center for Strategic and International Studies (CSIS), January 9, 2023, at https://www.csis.org/analysis/first-battle-next-war-wargaming-chinese-invasion-taiwan.
4. Jeffrey Lewis, David Joel La Boon, and Decker Eveleth, "China's Growing Missile Arsenal and the Risk of a 'Taiwan Missile Crisis,'" *Nuclear Threat Initiative*, Report November 18, 2020.
5. US Army, *Field Manual 3-0 Operations* (Washington, D.C.: Headquarters Department of the Army, October 2022), 7-4.

6. Israel Aerospace Industries (IAI), "HAROP Loitering Munition System," IAI website, accessed January 23, 2023, at https://www.iai.co.il/p/harop.
7. Unmanned aerial system (UAS) includes both the unmanned aerial vehicle (UAV) which flies and the control system, which usually involves a remote-control station which connects to the UAV by a radio link. A UAV is a component of a UAS.
8. BulgarianMilitary.com, "8 Russian 122mm Howitzers Destroyed by Ukrainian Drone Strikes," Bulgarian Government Publication, March 25, 2022, at https://youtu.be/4Erlh7JjNT0; and https://www.facebook.com/bulgarianmilitary/posts/watch-8-russian-122mm-howitzers-were-struck-after-uav-found-them-ukraine-ukraine/3047235708858925/; and https://www.zenger.news/2022/03/25/video-ukrainian-forces-take-out-russian-equipment-with-turkish-drones/.
9. Valius Venckunas, "Baykar drone factory in Ukraine to be complete in two years: CEO," *Aerotime Hub*, October 28, 2022, at www.aerotime.aero/articles/32524-baykar-factory-in-ukraine-to-be-complete-in-two-years-ceo.
10. Rohit Ranjan, "Russia Claims 2,911 Ukrainian Military Infrastructure Facilities Destroyed," *Republicworld.com*, March 10, 2022, at https://www.republicworld.com/world-news/russia-ukraine-crisis/russia-claims-2911-ukrainian-military-infrastructure-facilities-destroyed-article-show.html.
11. Roberta Morgan Wohlstetter, *Pearl Harbor, Warning and Decision* (Stanford: Stanford University Press, 1962), 382.
12. Ibid., 387.

Chapter 4

1. I paraphrase the quote "Untutored courage is useless in the face of educated bullets" which is attributed to Gen. George S. Patton, Jr. a general in the United States Army during World War II. Unfortunately, there is no record of this quote in any of Patton's speeches or writings, and it is not included in any of the official collections of his quotes or sayings. This statement means that bravery alone is not enough when facing an opponent who is well-prepared and well-armed, especially if those systems are modern precision weapons linked to a sensor network.
2. The Pantsir-S1 (NATO reporting name *Greyhound*) is a self-propelled, medium-range, surface-to-air missile and anti-aircraft artillery system, specifically designed to defeat aircraft, helicopters, precision munitions, cruise missiles, and drones.
3. Stijn Mitzer and Joost Oliemans, "Defending Ukraine – Listing Russian Military Equipment Destroyed By Bayraktar TB2s," *Oryx website*, February 27, 2022, at https://www.oryxspioenkop.com/2022/02/defending-ukraine-listing-russian-army.html; and Ukrainian Armed Forces, "Bayraktar destroyed the Russian Buk air defense system near Zhytomyr," Rubryka, February 27, 2022, at https://rubryka.com/en/2022/02/27/znyshhyv-rosijskyj-zrk-buk-pid-zhytomyrom/.
4. Dylan Malyasov, "Ukrainian Bayraktar TB2 Drones Successfully Attack Russian Convoys," *Defense Blog*, February 28, 2022, at https://defence-blog.com/ukrainian-bayraktar-tb2-drones-successfully-attack-russian-convoys/; and Lauren Kahn, "How Ukraine Is Using Drones Against Russia," Council of Foreign Relations (CFR), March 2, 2022, at https://www.cfr.org/in-brief/how-ukraine-using-drones-against-russia.
5. Uzi Rubin, "The Second Nagorno-Karabakh War: A Milestone in Military Affairs," The Begin-Sadat Center for Strategic Studies, Mideast Security and Policy Studies No. 184, December 2020, at https://besacenter.org/nagorno-karabakh-war-milestone/.

6. Leo Sands, "Sunken Russian warship *Moskva*: What do we know?" *BBC News*, April 18, 2022, at https://www.bbc.com/news/world-europe-61103927.
7. Stefano D'urso, "Evidence Of US-supplied Switchblade Loitering Munitions Targeting Russian Troops In Ukraine Emerges," *The Aviationist*, May 25, 2022, at https://theaviationist.com/2022/05/25/switchblade-ukraine/.
8. David Hambling, "Paper Planes? Ukraine Gets Flat-Packed Cardboard Drones From Australia," *Forbes*, March 6, 2023, at https://www.forbes.com/sites/davidhambling/2023/03/06/paper-planes-ukraine-gets-flat-packed-cardboard-drones-from-australia/?sh=2d2bafc1b8a2.
9. Felipe Dominguez "Raytheon Intelligence & Space to build mobile 50kW-class laser for US Army," *Raytheon News*, September 7, 202, at https://www.raytheonintelligenceandspace.com/news/2021/09/07/ris-build-mobile-50kw-class-laser-army.
10. Brett Tingley, "Jet-Powered Coyote Drone Defeats Swarm In Army Tests," *The War Zone*, thedrive.com, July 26, 2021, at https://www.thedrive.com/the-war-zone/41689/latest-coyote-drone-variant-defeats-drone-swarm-in-new-army-tests.
11. Can Kasapoglu, "Dangerous Drone for All Seasons: Assessing the Ukrainian Military's Use of the Bayraktar TB2," *Eurasia Daily Monitor*, Volume 19, Issue 36, March 16, 2022, at https://jamestown.org/program/a-dangerous-drone-for-all-seasons-assessing-the-ukrainian-militarys-use-of-the-bayraktar-tb-2/.
12. Benjamin Scott, "Army Counter-UAS, 2021–2028," *Military Review*, March–April 2021, at https://www.armyupress.army.mil/Portals/7/military-review/Archives/English/MA-21/Scott-Counter-UAS-1.pdf.

Chapter 5

1. Peter G. Tsouras, *Warriors' Words, A Quotation Book, From Sesostris to Schwarzkopf, 187 BC to AD 1991* (London: Arms and Armour Press, 1992), 405.
2. Joint Chiefs of Staff, Joint Publication 3-60, Joint Targeting, January 31, 2013 I–9. High-Value and High-Payoff Targets. An HVT is a target that the enemy commander requires for the successful completion of the mission. HPTs are derived from the list of HVTs. The loss of HVTs would be expected to seriously degrade important enemy functions throughout the friendly commander's area of interest. An HPT is one whose loss to the enemy will significantly contribute to the success of the friendly course of action.
3. "Israel Completes 'Iron Wall' Underground Gaza Barrier," *Al Jazeera*, December 7, 2021, at https://www.aljazeera.com/news/2021/12/7/israel-announces-completion-of-underground-gaza-border-barrier
4. Toi Staff, "Report: IAF bombing of Hamas 'Metro' Smashed Miles of Tunnels; No Info On Deaths," *The Times of Israel*, May 14, 2021, at https://www.timesofisrael.com/report-heavy-bombing-of-hamas-metro-destroyed-miles-of-tunnels-killed-dozens/.
5. Judah Ari Gross, "IDF Intelligence Hails Tactical Win in Gaza, Can't Say How Long Calm Will Last," *The Times of Israel*, May 27, 2021, at https://www.timesofisrael.com/idf-intel-hails-tactical-win-over-hamas-but-cant-say-how-long-calm-will-last/.
6. "Real-time processing requires a continual input, constant processing, and steady output of data, while near real-time processing is when speed is important but processing time in minutes is acceptable in lieu of seconds." Cristy Wilson, "The Difference Between Real-Time, Near Real-Time, and Batch Processing in Big Data," *Precisely*, November 14, 2022, at https://www.precisely.com/blog/big-data/difference-between-real-time-near-real-time-batch-processing-big-data.
7. Sebastien Roblin, "Israel's Bombardment Of Gaza: Methods, Weapons And Impact," *Forbes*, May 26, 2021, at https://www.forbes.com/sites/sebastienroblin/2021/05/26/israels-bombardment-of-gaza-methods-weapons-and-impact/?sh=1e5f0fdb2f44.

8. Avi Kalo, "AI-Enhanced Military Intelligence Warfare Precedent: Lessons from IDF's Operation "Guardian of the Walls," *Frost and Sullivan*, accessed March 7, 2023, at https://www.frost.com/frost-perspectives/ai-enhanced-military-intelligence-warfare-precedent-lessons-from-idfs-operation-guardian-of-the-walls/.
9. US Army, *Field Manual 3-0, Operations* (Washington D.C.: Headquarters Department of the Army, 2022), 3-3.
10. Ibid.
11. B. J. Copeland, "Artificial Intelligence," *Encyclopedia Britannica*, February 16, 2023, accessed March 6, 2023, at https://www.britannica.com/technology/artificial-intelligence.
12. James Chen, "What Is a Neural Network?" *Investopedia*, September 21, 2022, at https://www.investopedia.com/terms/n/neuralnetwork.asp.
13. Margaret Rouse, "Definition of Generative AI," *Technopedia*, February 1, 2023, at https://www.techopedia.com/definition/34633/generative-ai.
14. Open AI, ChatGPT App, accessed March 7, 2023, at https://chatgptonline.net/.
15. Tsouras, *Warrior's Words*, 434.
16. Christian Brose, "Testimony of Christian Brose to the HASC on Cyber, Information Technologies, and Innovation The Future of War: Is the Pentagon Prepared to Deter and Defeat America's Adversaries," Washington, D.C., US House of Representatives, Armed Services Committee, February 9, 2023, at https://armedservices.house.gov/hearings/citi-hearing-future-war-pentagon-prepared-deter-and-defeat-america-s-adversaries.
17. Stephen I. Schwartz, "The Costs of the Manhattan Project," Brookings Institute, accessed March 7, 2023, at https://www.brookings.edu/the-costs-of-the-manhattan-project/.
18. Translated by Flora Sapio (FLIAScholar), Weiming Chen and Adrian Lo (FLIA Research Intern) Foundation for Law and International Affairs, "New Generation of Artificial Intelligence Development Plan," Chinese Communist Party State Council Document No. 35, July 8, 2017, at https://flia.org/notice-state-council-issuing-new-generation-artificial-intelligence-development-plan/.
19. Deloitte Development LLC, "China Emerges as Global Tech, Innovation Leader," *Wall Street Journal*, 2019, at https://deloitte.wsj.com/articles/china-emerges-as-global-tech-innovation-leader-01572483727.
20. Jamie Gaida, Jennifer Wong-Leung, Stephan Robin and Danielle Cave, "ASPI's Critical Technology Tracker: The Global Race for Future Power," *The Australian Strategic Policy Institute Limited*, 2023, at https://ad-aspi.s3.ap-southeast-2.amazonaws.com/2023-03/ASPIs%20Critical%20Technology%20Tracker_0.pdf?VersionId=ndm5v4DRMfpLvu.x69Bi_VUdMVLp07jw, 1–2.

Chapter 6

1. Gregory C. Allen, "Understanding China's AI Strategy, Clues to Chinese Strategic Thinking on Artificial Intelligence and National Security," *Center for New American Security*, February 2019, at https://www.cnas.org/publications/reports/understanding-chinas-ai-strategy, 5–6.
2. Techmango, "ChatGPT & Dall-e-2: Everything You Need to Know About the Newest Passion," December 21, 2022, at https://www.techmango.net/chatgpt-dall-e-2-everything-you-need-to-know-about-the-newest-passion.
3. Cliff Saran, "Stanford University's AI Index 2019 annual report has found that the speed of artificial intelligence (AI) is outpacing Moore's Law," *Computerweekly.com*, December 12, 2019, at https://www.computerweekly.com/news/252475371/Stanford-University-finds-that-AI-is-outpacing-Moores-Law.
4. Chris Anderson, "Ray Kurzweil on What the Future Holds Next," TED2018 Interview, TED Conferences, LLC, 2018, at https://www.ted.com/talks/the_ted_interview_ray_kurzweil_on_what_the_future_holds_next/transcript.

5. Army Science Board, *Multidomain Operations, Final Report* (Washington, D.C.: Department of the Army, Office of the Deputy Under Secretary of the Army, May 2019), 8.
6. Sun Tzu, *The Art of War,* (Peter Harris, trans.) (London: Everyman's Library, 2018), 8.
7. "Combined arms is the synchronized and simultaneous application of arms to achieve an effect greater than if each element was used separately or sequentially." US Army ADP 3-0, Operations, Headquarters Department of the Army, October 2019, 3–9.
8. Army Doctrine Publication No. 6-0, *Mission Command: Command and Control of Army Forces* (Washington, D.C.: Headquarters Department of the Army, 2019). See the glossary for a definition of Distributed Mission Command.
9. David K. Spencer, Stephen Duncan, Adam Taliaferro, "Operationalizing artificial intelligence for multidomain operations: a first look," Proc. SPIE 11006, Artificial Intelligence and Machine Learning for Multidomain Operations Applications, May 10, 2019, at https://doi.org/10.1117/12.2524227.
10. Peter J. Boyer, "A Different War: Is the Army Becoming Irrelevant?" *The New Yorker*, June 23, 2002, at https://www.newyorker.com/magazine/2002/07/01/a-different-war.

Chapter 7

1. Ray Alderman, "Transitioning from the Kill Chain to the Kill Web," *Military Embedded Systems*, May 30, 2018, at https://militaryembedded.com/comms/communications/transitioning-from-the-kill-chain-to-the-kill-web.
2. "Ukraine says it hit Russian Command Post," *Associated Press News*, April 23, 2022, at https://apnews.com/article/russia-ukraine-kyiv-business-crimea-europe-6e0de7b8a92ef0af3c60b69438d6b8b1.
3. John M. Doyle, "General: Precise Sensors to Close Kill Chain is a Key Takeaway from Ukraine War," *Seapower, The Official Publication of the Navy League of the United States*, May 5, 2022.
4. Mara Karlin, "The Kill Chain: An Interview with Christian Brose," Johns Hopkins University, School of Advanced International Studies, May 26, 2020, at https://sais.jhu.edu/news-press/event-recap/kill-chain-interview-christian-brose; and Christian Brose, *The Kill Chain: Defending America in the Future of High-Tech Warfare* (New York: Hachette Books, 2020).
5. Sydney J. Freedberg, "Target Gone In 20 Seconds: Army Sensor-Shooter Test," *Speaking Defense*, September 10, 2020, at https://breakingdefense.com/2020/09/target-gone-in-20-seconds-army-sensor-shooter-test/.
6. David H. Freedman, "US Is Only Nation with Ethical Standards for AI Weapons. Should We Be Afraid?" *Newsweek*, September 15, 2021, at https://www.newsweek.com/2021/09/24/us-only-nation-ethical-standards-ai-weapons-should-we-afraid-1628986.html.
7. Training and Doctrine Command, *ATP 7-100.3 Chinese Tactics* (Washington D.C.: US Army, 2021), 1–10.
8. Alderman, "Transitioning from the Kill Chain to the Kill Web."
9. Avi Kalo, "AI-Enhanced Military Intelligence Warfare Precedent: Lessons from IDF's Operation 'Guardian of the Walls,'" *Frost and Sullivan*, July 9, 2021, at https://www.frost.com/frost-perspectives/ai-enhanced-military-intelligence-warfare-precedent-lessons-from-idfs-operation-guardian-of-the-walls/.
10. Anna Ahronheim, "Israel's Operation Against Hamas was the World's First AI War," *The Jerusalem Post*, May 27, 2021.
11. Ibid.
12. Robin Laird and Edward Timberlake, *A Maritime Kill Web Force in the Making: Deterrence and Warfighting in the XXIst Century* (Pennsauken: Bookbaby, 2022).

Chapter 8

1. David Hambling, "The next era of drones will be defined by 'swarms'," *BBC, Future Now*, April 16. 2017, at https://www.bbc.com/future/article/20170425-were-entering-the-next-era-of-drones.
2. See dramatic video of the attack, from the perspective of one of the Ukrainian USVs that conducted the attack at https://twitter.com/i/status/1586460767619977216; and H.I Sutton, "Why Ukraine's Remarkable Attack On Sevastopol Will Go Down In History," November 17, 2022, at https://www.navalnews.com/naval-news/2022/11/why-ukraines-remarkable-attack-on-sevastopol-will-go-down-in-history/.
3. John Arquilla and David Ronfeldt, *Swarming and the Future of Conflict* (Santa Monica: RAND Corporation. 2000), at https://www.rand.org/pubs/documented_briefings/DB311.html, vii.
4. Sean J. A. Edwards, *Swarming and the Future of Warfare* (Santa Monica: RAND Corporation, 2005), at https://www.rand.org/pubs/rgs_dissertations/RGSD189.html, xxii.
5. Ibid., xvii.
6. Joseph Trevithick, "China Conducts Test Of Massive Suicide Drone Swarm Launched From A Box On A Truck," *The War Zone*, thedrive.com, October 14, 2020, at https://www.thedrive.com/the-war-zone/37062/china-conducts-test-of-massive-suicide-drone-swarm-launched-from-a-box-on-a-truck.
7. Christian Brose, "Testimony of Christian Brose To the House Armed Services Committee Subcommittee on Cyber, Information Technologies, and Innovation The Future of War: Is the Pentagon Prepared to Deter and Defeat America's Adversaries," Washington, D.C.: US House of Representatives, Armed Services Committee, February 9, 2023, at https://armedservices.house.gov/hearings/citi-hearing-future-war-pentagon-prepared-deter-and-defeat-america-s-adversaries
8. Insider Business, "Watch the Navy's LOCUST (Low-Cost UAV Swarming Technology), Launcher Fire a Swarm of Drones," *YouTube*, April 20, 2017, at https://www.youtube.com/watch?v=qW77hVqux10.
9. The Coyote is a small, expendable loitering munition made by Raytheon Corporation.
10. Joseph Trevithick, "Massive Drone Swarm Over Strait Decisive In Taiwan Conflict Wargames: Air Force and Independent Think Tank Simulations Show Giant Drone Swarms are Key to Defeating China's Invasion of Taiwan," *The War Zone*, thedrive.com, May 19, 2022, at https://www.thedrive.com/the-war-zone/massive-drone-swarm-over-strait-decisive-in-taiwan-conflict-wargames.
11. US Department of Defense, "Defense Department Successfully Transitions New Technology to Programs of Record," March 16, 2021, at https://www.defense.gov/News/Releases/Release/Article/2539182/defense-department-successfully-transitions-new-technology-to-programs-of-record/.
12. David Hambling, *Swarm Troopers: How Small Drones Will Conquer the World* (Archangel Ink, 2015), 8.
13. Tyler Rogoway as quoted in an article by Joseph Trevithick. "Army Buys Small Suicide Drones To Break Up Hostile Swarms And Potentially More." *The War Zone, The Drive*, Jul. 17, 2018.

Chapter 9

1. Sydney J. Freedberg, "War Without Fear: DepSecDef Work on How AI Changes Conflict," *Breaking Defense*, May 31, 2017, at https://breakingdefense.com/2017/05/killer-robots-arent-the-problem-its-unpredictable-ai/.

2. For more on the Battle of Cowpens see "Chapter 6, Charge Bayonets! Daniel Morgan at the Battle of Cowpens," in John Antal's *7 Leadership Lessons of the American Revolution* (London: Casemate, 2013).
3. Michael Howard and Peter Paret trans. *On War, Carl von Clausewitz* (Princeton: Princeton University Press, 1989), 578.
4. For more on the OODA Loop see: Robert Coram, *Boyd: The Fighter Pilot Who Changed the Art of War* (New York: Back Bay Books, 2004). See Glossary for definition of an ADCOP. Also see "Patterns of Conflict" at https://youtu.be/lzRqZnPVeJI.
5. Author's definition.
6. Gareth Halfacree, "A New Non-Invasive Brain-Machine Interface Offers Thought-Based Robot Control with High Accuracy," *Hackster*, March 24, 2023, at https://www.hackster.io/news/a-new-non-invasive-brain-machine-interface-offers-thought-based-robot-control-with-high-accuracy-825e3d406d62.
7. See Johns Hopkins University COVID dashboard at https:// coronavirus.jhu.edu/map.html.
8. MMO Populations, "World of Tanks Player Count, (Rank 21/138 of all MMOs)," *mmo.population.com*, as of January 24, 2023, at https://mmo-population.com/r/worldoftanks/.
9. A diegetic view occurs within the world of a narrative rather than as something external to that world. The story is told or recounted, as opposed to shown or enacted. For instance, in a film, music is coming from a car radio in the scene rather than an unseen orchestra.
10. To see the minimalist approach to the Metroid Heads-Up Display in the Metroid video game, see https://metroid.fandom.com/wiki/Heads-Up_Display.
11. MetaQuotes, "MetaTrader 5 Trading Platform," *metaquotes.net*, January 23, 2023, at https://www.metaquotes.net/en/metatrader5.
12. Alexander Suvorov, "The Art of Victory," English trans. 1800, at https://annas-archive.org/md5/12e0a1ed0963b42f539b67e29e983f32.
13. John R. Hoehn, "Joint All Domain Command and Control (JADC2), Congressional Research Service Report (CRS), July 1, 2021," Congress.gov, March 18, 2021, at https://crsreports.congress.gov/product/pdf/R/R46725/2, 1.
14. Ibid., 2.
15. Ibid., 11.
16. Erik Brynjolfsson, coauthor of "The Second Machine Age: Work, Progress, and Prosperity in a Time of Brilliant Technologies," LinkedIn Speaker Series talk, February 22, 2018, at https://speakerseries.libsyn.com/webpage/2018/02/22.
17. Hoehn, "Joint All Domain Command and Control," 4.
18. Paul Scharre, *Centaur Warfighting, The False Choice of Humans vs. Automation* (Temple International and Company L.J., 2016), 152.
19. Hoehn, "Joint All Domain Command and Control," 4.
20. Sgt. Joshua Oh, "Distributed C2 Concept to Address the Pacific Theater," US Army, October 17, 2022, at https://www.army.mil/article/261211/distributed_c2_concept_to_address_the_pacific_theater.
21. Shaikh Nayeem Faisal, et al. "Noninvasive Sensors for Brain–Machine Interfaces Based on Micropatterned Epitaxial Graphene," American Chemical Society (ACS), March 16, 2023, at https://pubs.acs.org/doi/10.1021/acsanm.2c05546.

Chapter 10

1. Sydney J. Freedberg Jr., "Army's New Aim Is 'Decision Dominance,'" *Breaking Defense*, March 17, 2022, at https://breakingdefense.com/2021/03/armys-new-aim-is-decision-dominance/.
2. Bevin Alexander, *How Great Generals Win* (New York: Avon Books, 1993), 21.

3. Merrick Krause, "Decision Dominance: Exploiting Transformational Asymmetries," *Defense Horizons*, The Center for Technology and National Security Policy National Defense University, February 2003, at https://ndupress.ndu.edu/Portals/68/Documents/defensehorizon/DH-023.pdf?ver=2016-11-15-092812-353
4. Freedberg, "Army's New Aim Is 'Decision Dominance.'"
5. Matt Gonzales, "Mobile Satellite System Reduces Communications Gaps, Increases Naval Interoperability," *US Marine Corps Systems Command*, December 15, 2021, at https://www.marines.mil/DesktopModules/ArticleCS/Print.aspx?PortalId=1&ModuleId=632&Article=2874262.
6. US Army, *FM 3.0 Operations* (Washington, D.C.: Headquarters Department of the Army, 2022), 3–14.
7. Thomas Cleary trans. *The Art of War, Complete Texts and Commentaries, Sun Tzu* (Boston: Shambhala Publications Inc., 2003), 73.
8. See https://vbs4.com/.
9. Eric Limer, "The Norwegian Army is Using the Oculus Rift to See Through Its Tanks," *Gizmodo*, May 5, 2014, at https://gizmodo.com/the-norwegian-army-is-using-the-oculus-rift-to-see-thro-1571831534.

Chapter 11

1. John Hoist, "Starlink, China, and Upset Plans," *Ill-Defined Space*, May 12, 2022, at https://illdefinedspace.substack.com/p/starlink-china-and-upset-plans.
2. Qiao Liang and Wang Xiangsui, *Unrestricted Warfare* (Beijing: PLA Literature and Arts Publishing House, 1999), at https://www.c4i.org/unrestricted.pdf, 11.
3. Ashish Dangwal, "China Sends Suspected Military Reconnaissance Balloons Over Taiwan Amid Russian Ops In Ukraine; Beijing Responds," *The EurAsian Times*, March 2, 2022, at and https://eurasiantimes.com/china-launch-military-balloons-into-taiwan-amid-ukraine/; and Qiang Tianlin, "The 'Sharp Weapon' of Air-Space Battle, *China Military Network*, Ministry of Defense Network, March 30, 2018, at http://www.81.cn/jfjbmap/content/2018-03/30/content_202810.htm.
4. Iain Boyd, "Chinese spy balloon over the US: An Aerospace Expert Explains How the Balloons Work and What They Can See," *The Conversation*, February 4, 2023, at https://theconversation.com/chinese-spy-balloon-over-the-us-an-aerospace-expert-explains-how-the-balloons-work-and-what-they-can-see-199245.
5. Heather Chen and Wayne Chang, "China Says It 'Reserves The Right' To Deal With 'Similar Situations' After US Jets Shoot Down Suspected Spy Balloon," *CNN*, February 5, 2023, https://www.cnn.com/2023/02/04/asia/beijing-reacts-us-jets-shoot-chinese-spy-balloon-intl-hnk/index.html.
6. Fatima Khaled, "Ukraine Official Asks Elon Musk for Starlink Stations Amid Russian Invasion," *Newsweek*, February 26, 2022, at https://www.newsweek.com/ukraine-official-asks-elon-musk-starlink-stations-amid-russian-invasion-1682977.
7. Lexi Lonas, "Zelensky says Ukraine Receiving More SpaceX Internet Stations for 'Destroyed Cities,'" *The Hill*, Mar. 5, 2022, atchhttps://thehill.com/policy/international/597028-zelensky-says-ukraine-receiving-more-spacex-internet-stations-for/.
8. Elon Musk @elonmusk Twitter. 1:49 pm, March 3, 2022, at https://twitter.com/elonmusk/status/1499472139333746691.
9. Erick Mack, "US Military Says SpaceX Handily Fought Off Russian Starlink Jamming Attempts," *CNET*, April 22, 2022, at https://www.cnet.com/science/space/us-military-says-spacex-handily-fought-off-russian-starlink-jamming-attempts/.

10. Alia Shoaib, "An Elite Ukrainian Drone Unit Exploits the Cover of Night to Destroy Russian Tanks and Trucks While Their Soldiers Sleep," *Business Insider*, March 20, 2022, at https://www.businessinsider.com/ukrainian-drone-unit-strikes-russian-targets-while-they-sleep-the-times-2022-3?op=1.
11. Tom Simonite, "How Starlink Scrambled to Keep Ukraine Online: Elon Musk's intervention demonstrates how satellite internet could route around war or censorship far beyond Ukraine," Wired, May 11, 2022, at https://www.wired.com/story/starlink-ukraine-internet/.
12. Ilya TZsukanov, "Chinese Military 'Deeply Alarmed' Over Musk's Starlink Satellites' Dual-Use Capabilities," Sputnik International, May 11, 2022, at https://sputniknews.com/20220511/chinese-military-deeply-alarmed-musks-starlink-satellites-dual-use-capabilities-1095443590.html.
13. David Cowhig, "PRC Defense: Starlink Countermeasures," David Cowhig's Translation Blog, May 25, 2022, at https://gaodawei.wordpress.com/.
14. Brandon Wall and Nicholas Ayrton, "Drones and Starlink: Combining Satellite Constellations with Unmanned Navy Ships," *Center for International Maritime Security*, September 1, 2021, at https://cimsec.org/drones-and-starlink-combining-satellite-constellations-with-unmanned-navy-ships/.
15. Mike Wall, "SpaceX launches 56 Starlink Satellites, Lands Rocket on Ship at Sea," *Space.com*, March 24, 2023, at https://www.space.com/spacex-starlink-satellites-group-5-5-launch.
16. Michael Kan, "Starlink on a Drone? This Company Is Working on the Idea," *PCMag.com*, November 2, 2022, at https://www.pcmag.com/news/starlink-on-a-drone-this-company-is-working-on-the-idea.
17. SpaceX website, "Starshield," at SpaceX.com, accessed March 28, 2023, at https://www.spacex.com/starshield/.

Chapter 12

1. Margarita Konaev and Kirstin J.H. Brathwaite, "Russia's Urban Warfare Predictably Struggles Fighting in Cities is Hard for any Military," *Forbes Magazine*, April 4, 2022, at https://foreignpolicy.com/2022/04/04/russia-ukraine-urban-warfare-kyiv-mariupol/.
2. "House to House Fighting in Bakhmut," Military Mind, *TVP World*, Polish Broadcasting Service, December 12, 2022; and "Intense urban combat in Bakhmut," Military Mind, *TVP World*, Polish Broadcasting Service, December 16, 2022, at https://www.youtube.com/watch?v=bv96p2f5APQ and https://www.youtube.com/watch?v=6nPBsWwvLBY.
3. *Rasputitsa* is the name for the wet, muddy period caused by melting snow in the spring and heavy rains in autumn in Ukraine and Russia. *Rasputitsa* is famous for slowing and hampering the movement of armies as it did for Napoleon in 1812 and the German Army during World War II. *Bezdorizhzhia*, the Ukrainian name for *Rasputitsa*, translates as "roadlessness" where off-road movement is extremely difficult or impossible due to the mud.
4. Sun Tzu said, "The lowest (strategy) is to attack a city. Siege of a city is only done as a last resort." Thomas Cleary trans. *The Art of War, Complete Texts and Commentaries, Sun Tzu* (Boston: Shambhala Publications Inc., 2003), 73.
5. Marc Harris, et al. "Megacities and the United States Army: Preparing for a Complex and Uncertain Future," Chief of Staff of the Army Strategic Studies Group, June 2014, 3.
6. For an excellent discussion about tanks and modern warfare, see David Johnson, "The Tank is Dead: Long Live the Javelin, the Switchblade, the …?" *War on the Rocks*, April 18, 2022, at https://warontherocks.com/2022/04/the-tank-is-dead-long-live-the-javelin-the-switchblade-the/.
7. Elliot Gardner, "All Clear: the Onset of See-Through Armour," *Army Technology*, August 28, 2018, at https://www.army-technology.com/features/clear-onset-see-armour/.
8. "M5 Ripsaw, Extreme Mobility, Decisive Lethality, Wingman Ready," *Team Ripsaw*, Textron Systems, accessed March 2, 2023, at https://www.textronsystems.com/products/ripsaw-m5.

9. Tyler Rogoway, "AbramsX Next Generation Main Battle Tank Breaks Cover," *The War Zone*, thedrive.com, October 9, 2022, at https://www.thedrive.com/the-war-zone/abramsx-next-generation-main-battle-tank-breaks-cover.
10. Mark Episkopos, "Russia Is Ready to Receive Shiny New T-14 Armata Tanks," *The National Interest*, July 7, 2021, at https://nationalinterest.org/blog/buzz/russia-ready-receive-shiny-new-t-14-armata-tanks-189320.
11. Edge computing is a decentralized computing infrastructure where the processing, storage, and analysis of data takes place in or near sensors, devices, and terminals that collect and generate data, rather than sending it all to a server or to a central cloud. See https://www.ibm.com/cloud/what-is-edge-computing.
12. For a review of the Battle of Mariupol, see Peter Beaumont, "Defenders of Mariupol are the Heroes of Our Time: the Battle that Gripped the World," *The Guardian*, May 17, 2022, at https://www.theguardian.com/world/2022/may/17/defenders-of-mariupol-are-the-heroes-of-our-time-the-battle-that-gripped-the-world.
13. "Legion-X, Multidomain Autonomous Network Combat Solutions for Unmanned Heterogeneous Swarms," *Elbit Systems*, accessed March 2, 2023, at https://elbitsystems.com/product/legion-x/.
14. "Lanius, Drone-based Loitering Munition for Complex Environments," *Elbit Systems*, accessed March 2, 2023, at https://elbitsystems.com/product/lanius/.
15. Dr. Matt Turek, "Explainable Artificial Intelligence (XAI)," Defense Advanced Research Projects Agency (DARPA), accessed March 2, 2023, at https://www.darpa.mil/program/explainable-artificial-intelligence.
16. Joe Saballa, "US Air Force Tactical Fighter Flown by Artificial Intelligence for First Time," *The Defense Post*, February 14, 2023, at https://www.thedefensepost.com/2023/02/14/us-fighter-artificial-intelligence/.
17. "OFFensive Swarm-Enabled Tactics (OFFSET)," Defense Advanced Research Projects Agency (DARPA), accessed March 2, 2023, at https://www.darpa.mil/work-with-us/offensive-swarm-enabled-tactics.
18. John Spires, "DJI Named Top Commercial Drone Maker with 70% Market Share," *DroneDJ*, October 16, 2020, at https://dronedj.com/2020/10/16/dji-named-top-commercial-drone-maker-with-70-market-share/.
19. Alia Shoaib, "Inside the elite Ukrainian Drone Unit Founded by Volunteer IT Experts: 'We are all soldiers now,'" *Business Insider*, April 9, 2022, at https://www.businessinsider.com/inside-the-elite-ukrainian-drone-unit-volunteer-it-experts-2022-4.
20. Mandip Singh, "Global MALE & HALE UAV Key Developments Across Global Top 10 Defense Spenders," *European Security and Defense*, January 10, 2023., at https://euro-sd.com/2023/01/articles/26779/global-male-hale-uav-key-developments-across-global-top-10-defence-spenders/.
21. Carlo Munoz, "US Navy, CENTCOM Seek Solutions for Stratospheric ISR Operations," *Janes*, April 5, 2021, at https://www.janes.com/defence-news/news-detail/us-navy-centcom-seek-solutions-for-stratospheric-isr-operations.
22. Lee Ferran, "Army's Ultra-Endurance Zephyr Drone Comes Down After 'Unexpected Termination' over Arizona Desert," *Breaking Defense*, August 23, 2022, at https://breakingdefense.com/2022/08/armys-ultra-endurance-zephyr-drone-comes-down-after-unexpected-termination-over-arizona-desert/.
23. W. J. Hennigan, "Downed Chinese Balloon Part of Global Spy Operation, Pentagon Alleges," *Time*, February 8, 2023, https://time.com/6253974/chinese-balloon-worldwide-spy-operation/.

24. "Sentinels of the Sky: The Persistent Threat Detection System," Lockheed Martin, accessed March 2, 2023, at https://www.lockheedmartin.com/en-us/news/features/history/ptds.html.
25. Paul Scharre, "Robotics on the Battlefield Part II: The Coming Swarm," Center for a New American Security (CNAS), October 15, 2014, at https://www.cnas.org/publications/reports/robotics-on-the-battlefield-part-ii-the-coming-swarm.

Chapter 13

1. Gen. Kenneth F. McKenzie, Jr., "Posture Statement of US Central Command before the Senate Armed Services Committee," April 22, 2021, at https://www.armed-services.senate.gov/imo/media/doc/McKenzie%20Testimony%2004.22.211.pdf.
2. Sébastien Roblin, "In 1944, USS *Intrepid* Survived Assault from a Huge Japanese Warship," *The National Interest*, June 6, 2021, at https://nationalinterest.org/blog/reboot/1944-uss-intrepid-survived-assault-huge-japanese-warship-186877.
3. Steve Balestrieri, "Remembering Adm. John Thach, Naval Aviator, Died on This Day 1981," *SOFREP*, April 15, 2019, at https://sofrep.com/specialoperations/remembering-adm-john-thach-naval-aviator-died-on-this-day-1981/.
4. Mike Benitez, "OA-X: More Than Just Light Attack," *War on the Rocks*, August 16, 2016, at https://warontherocks.com/2016/08/oa-x-more-than-just-light-attack/.
5. Stephen Losey, "US Special Operations Command chooses L3Harris' Sky Warden for Armed Overwatch Effort," Defense News, August 1, 2022, at https://www.defensenews.com/air/2022/08/01/us-special-operations-command-chooses-l3harris-sky-warden-for-armed-overwatch-effort/.
6. BAE Systems transforms 2.75-in rockets, used in the US military for over 50 years, into precision counter-drone killers with a missile range of three kilometers. The APKWS system is the most cost-effective laser-guided munition in its class.

Chapter 14

1. Captain Wayne P. Hughes Jr., US Navy (retired), *Fleet Tactics and Coastal Combat, second edition* (Annapolis: Naval Institute Press, 2000), 4–5.
2. Congressional Research Service, "The Army's Optionally Manned Fighting Vehicle (OMFV) Program: Background and Issues for Congress," February 22, 2019, at https://crsreports.congress.gov/product/pdf/R/R45519/1.
3. Eric Tegler, "An Army General Says The Robotic Combat Vehicles It's Experimenting With Will Be The 'Ghosts Of Patton's Army,'" *Forbes Magazine*, July 30, 2021.
4. John Antal, *Leadership Rising: Raise Your Awareness, Raise Your Leadership, Raise Your Life* (Oxford, UK: Casemate Publishers, 2021), 129.
5. Ellie Cook, "How Russia's 'Marker' Combat Robots Could Impact Ukraine War," *Newsweek*, January 18, 2023, at https://www.newsweek.com/russia-marker-combat-robots-ukraine-tests-impact-1774666.

Chapter 15

1. General Douglas MacArthur, "MacArthur Endorses More Aid to Britain; General Cables Warning Against 'Fatal Epitaph, Too Late,'" *The New York Times*, September 16, 1940, at https://www.nytimes.com/1940/09/16/archives/macarthur-endorses-more-aid-to-britain-general-cables-warning.html.

2. David Axe, "The Ukrainians Keep Blowing Up Russian Command Posts And Killing Generals," *Forbes*, April 23, 2022, at https://www.forbes.com/sites/davidaxe/2022/04/23/the-ukrainians-keep-blowing-up-russian-command-posts-and-killing-generals/?sh=7b62b207a350.
3. Milford Beagle (LTG US Army), Jason Slider (BG US Army), and Matthew R. Arrol (LTC US Army), "The Graveyard of Command Posts What Chornobaivka Should Teach Us about Command and Control in Large-Scale Combat Operations," *Military Review Online*, March 2023, at https://www.armyupress.army.mil/Portals/7/military-review/Archives/English/Online-Exclusive/2023/Graveyard-of-Command-Posts/The-Graveyard-of-Command-Posts-UA2.pdf.
4. Sydney J. Freedberg Jr., "Firepower & People: Army Chief On Keys to Future War," *Breaking Defense*, October 10, 2022, at https://breakingdefense.com/2022/10/firepower-people-army-chief-on-keys-to-future-war-exclusive/?_hsmi=229138687&_hsenc=p2ANqtz-8JhaqIK0drCqwhSINGHWTz2nbsyRG3if2tFlE0PTmML6iIDJ2lFkWhgNd3D92s-kkfCGOR8YCpqYlinUtjKgd9zC7Icw.
5. Army Doctrine Publication No. 6-0, *Mission Command: Command and Control of Army Forces* (Washington, D.C.: Headquarters Department of the Army, 2019), at https://usacac.army.mil/node/2425, 1–3.
6. J. F. C. Fuller, *Generalship: Its Diseases and Their Cure: A Study of The Personal Factor in Command* (Harrisburg: Military Service Publishing Company, 1936), 61.
7. Andrew Eversden, "Inside the Army's Distributed Mission Command Experiments In, And Over, The Pacific," *Breaking Defense*, March 23, 2022, at https://breakingdefense.com/2022/03/inside-the-armys-distributed-mission-command-experiments-in-and-over-the-pacific/.
8. Beagle, Slider, and Arrol, "The Graveyard of Command Posts."
9. Jeremy Hofstetter and Adam Wojciechowski, "Electromagnetic Spectrum Survivability in Large-Scale Combat Operations," Fort Benning, GA: Infantry, Winter 2020–2021, at https://www.benning.army.mil/infantry/magazine/issues/2020/Winter/pdf/7_Hofstetter_EW.pdf, 22.
10. Hofstetter and Wojciechowski, "Electromagnetic Spectrum Survivability in Large-Scale Combat Operations," 23.
11. Lt. Gen. Sergei Sevryukov, "Statement by First Deputy Chief of Main Military-Political Directorate of Russian Armed Forces Lieutenant General Sergei Sevryukov," *mil.ru*, Russian Ministry of Defense Video, January 4, 2022.
12. Ibid., 5. Video has a page number?
13. An expando-van is a truck that can be expanded into a larger working space for command posts. See https://www.army.mil/article/219567/new_army_vehicles_being_developed_to_counter_modern_threats.
14. TRADOC G2. *The Red Team Handbook, The Army's Guide to Making Better Decisions*, US Army Training and Doctrine Command, Version 9. 2019, at https://usacac.army.mil/sites/default/files/documents/ufmcs/The_Red_Team_Handbook.pdf.
15. Lewis Sorley, *Press On! Selected Works of General Donn A. Starry Volume I* (Fort Leavenworth: US Army Combined Arms Center, Combat Studies Institute, 2009), 164.
16. An example of Mission Command on the Move can be found in an article by MAJ Adam R. Brady, LTC Tommy L. Cardone and CPT Edwin C. den Harder, "Mission Command on the Move," *Armor Magazine*, 2015, at https://www.benning.army.mil/armor/eARMOR/content/issues/2015/OCT_DEC/Brady-Cardone-Den-Harder.pdf.
17. Andrew Eversden, "Inside the Army's Distributed Mission Command Experiments In, and Over, the Pacific," *Breaking Defense*, March 23, 2022, at https://breakingdefense.com/2022/03/inside-the-armys-distributed-mission-command-experiments-in-and-over-the-pacific/.
18. David Axe, "To Hide From Ukraine's Drones, Russian Troops Could Lay Smoke Screens," *Forbes*, December 17, 2021, at https://www.forbes.com/sites/davidaxe/2021/12/17/to-hide-from-ukraines-drones-russian-troops-could-lay-smoke-screens/?sh=638eee9168ef.

19. Eversden, "Inside The Army's Distributed Mission Command Experiments In, And Over, The Pacific."
20. US Army, *FM 3.0 Operations*, Headquarters Department of the Army, Washington, D.C.: October 1, 2022, 3-2.

Chapter 16

1. Lewis Sorley, *Press On! Selected Works of General Donn A. Starry Volume I* (Fort Leavenworth: US Army Combined Arms Center, Combat Studies Institute, 2009), 249.
2. Multidomain Operations is visualized in six physical areas: the strategic support area, the operational support area, the tactical support area, the close area, deep maneuver area, and the deep fires area. See US Army, *Field Manual 3-0, Operations* (Washington, D.C.: Headquarters Department of the Army, 2022), 1–2.
3. Distributed Mission Command is an emerging concept concerning the execution of Mission Command in a mesh CP configuration that divides the conventional CP infrastructure into resilient "functional nodes" that are small, mobile, and dispersed throughout the battlespace, yet remain in constant communications. The purpose of Distributed Mission Command is to make the C4ISR nodes more resilient, survivable, and effective. (Author's definition)
4. Jared Keller, "The Army is Testing a Robot Mini-Tank Straight Out of 'Fast and the Furious' The Ripsaw is Here and Ready to Rock," *Task and Purpose*, July 27, 2021, at https://taskandpurpose.com/news/army-robotic-combat-vehicle-medium-ripsaw-testing/ and https://youtu.be/Wc_ChwLMgCY.
5. Kris Osborn, "Army Evaluates Newly Unveiled AbramsX Main Battle Tank for Future War Into 2050 General Dynamics Land Systems Unveils Breakthrough Abrams for Future," *Warrior Maven*, October 11, 2022, at https://warriormaven.com/land/army-evalutes-newly-unveiled-abrams-x-main-battle-tank-variant-for-future-warand https://youtu.be/TcfuyyxFtgQ.
6. Microsoft defines edge computing as the "means to allows devices in remote locations to process data at the 'edge' of the network, either by the device or a local server. And when data needs to be processed in the central datacenter, only the most important data is transmitted, thereby minimizing latency." See https://azure.microsoft.com/en-us/resources/cloud-computing-dictionary/what-is-edge-computing/.
7. J. F. C. Fuller, *Generalship, Its Diseases and Their Cure, A Study of the Personal Factor in Command* (Harrisburg: Military Service Publishing Co., 1936), at https://ia801606.us.archive.org/31/items/GeneralshipItsDiseasesAndTheirCure/GeneralshipItsDiseasesAndTheirCure.pdf.
8. Microsoft definition of edge computing. See https://azure.microsoft.com/en-us/resources/cloud-computing-dictionary/what-is-edge-computing/.
9. This is an essential tool to executing all domain operations to allow commanders to execute Distributed mission command. The power of Distributed Mission Command is its ability to join otherwise disconnected and independent units, thus increasing the resilience of the force. Distributed Mission Command can be defined as the conditional, adaptive delegation or assumption of Mission Command activities through orders or protocols to synchronize operations, maintain initiative, and achieve the commander's intent. Key to this effort is the ability of commanders to visualize the battlespace in real-time with an ADCOP.
10. US Army, *Field Manual 3-0, Operations* (Washington, D.C.: Headquarters Department of the Army, 2022), 1–7.
11. William Yang, "China: Fears Grow for Detained Anti-COVID Protesters," *Deutsche Welle*, January 17, 2023, at https://www.dw.com/en/china-fears-grow-for-detained-anti-covid-protesters/a-64421018; and Katherine Miller, "Many participants of the 'White Paper

Movement' arrested and lost contact," *TheBL*, December 6, 2022, at https://thebl.tv/china/many-participants-of-the-white-paper-movement-arrested-and-lost-contact.html and https://www.asianews.it/news-en/‘White-sheet’-revolution%27:-scores-of-Chinese-protesters-in-prison-57559.html, and https://globalvoices.org/2023/01/28/anti-zero-covid-white-paper-protesters-face-forced-disappearance-in-china/, and https://www.hrw.org/news/2023/01/26/china-free-white-paper-protesters.

12. 60 Minutes Overtime, "The Persistent Threat of China Invading Taiwan," *CBS News*, October 9, 2022. Admiral Lee Hsi-min, who once led Taiwan's armed forces, told correspondent Lesley Stahl about China, at https://www.cbsnews.com/news/taiwan-china-military-threat-60-minutes-2022-10-09/.

The methods of war are changing. Leaders must understand these changes and gain foresight to plan, prepare, and execute successful military operations. In this photo, the Ghost Robotics' Vision 60Q, a quadrupedal ground robot that is capable of maneuvering through rough terrain, conducts a reconnaissance mission at Holloman Air Force Base, New Mexico, on April 17, 2023. See: https://www.ghostrobotics.io/vision-60 (US Air Force photo by Airman 1st Class Isaiah Pedrazzini)

Glossary

ADCOP (All Domain Common Operational Picture). An ADCOP is a Common Operational Picture (COP) that enables a warfighter to visualize actions and relevant information in the battlespace in all five domains (land, sea, air, space, and cyber) and is shared by all pertinent commands automatically and in real-time. (Author's definition)

Artificial Intelligence. "… the term 'artificial intelligence' includes the following: (1) Any artificial system that performs tasks under varying and unpredictable circumstances without significant human oversight, or that can learn from experience and improve performance when exposed to data sets. (2) An artificial system developed in computer software, physical hardware, or other context that solves tasks requiring human-like perception, cognition, planning, learning, communication, or physical action. (3) An artificial system designed to think or act like a human, including cognitive architectures and neural networks. (4) A set of techniques, including machine learning, that is designed to approximate a cognitive task. (5) An artificial system designed to act rationally, including an intelligent software agent or embodied robot that achieves goals using perception, planning, reasoning, learning, communicating, decision making, and acting." US House of Representatives, "John S. McCain National Defense Authorization Act for Fiscal Year 2019, Conference Report to Accompany H.R. 5515," US Government Publishing Office, Washington D.C., July 25, 2018, at https://www.congress.gov/115/crpt/hrpt874/CRPT-115hrpt874.pdf, 64.

Battleshock. The operational, informational, and organizational paralysis induced by the rapid convergence of key disrupters in the battlespace. Battleshock occurs when the tempo of operations is so fast, and the multidomain means so overwhelming, that the enemy cannot think, decide, and act in time. (Author's definition)

Commander's Intent. A clear and concise expression of the purpose of the operation and the desired military end state that supports Mission Command, provides focus to the staff, and helps subordinate and supporting commanders act to achieve the commander's desired results without further orders, even when the operation does not unfold as planned. (JP 3-0)

Common Operational Picture (COP). "… a single identical display of relevant information shared by more than one command … A common operational picture facilitates collaborative planning and assists all echelons to achieve situational awareness." A COP is a command-and-control tool that provides situational awareness, enabling users to make accurate, informed decisions by integrating data from multiple sources into one spatial data platform. During the Battle of Britain during the summer of 1940, the British used an analog COP to track the air battles against the German air force. Named the "Dowding System," after Air Chief Marshal Sir H. C. T. Dowding, it involved a complex infrastructure of detection, command, and control that ran the battle on a large board where designated markers were moved by members of the Women's Auxiliary Air Force. Today, an analog COP consists of the critical information occurring in the battlespace and is depicted on a two-dimensional map and on supporting information boards or screens. A digital COP puts all this information on a digital screen where information can be input in real-time, in some cases automatically, to depict current actions in the battlespace.

Distributed Mission Command. The execution of Mission Command using smaller distributed command nodes to execute the functions of the command post without staff co-location. The goal of distributed mission command is to enhance continuity and survivability of the command function in a transparent and lethal battlespace. (Author's definition)

Internet of Battlespace Things (IoBT). Often referred to as the Internet of Battlefield Things, using the old term "battlefield" instead of "battlespace," IoBT is the concept of connecting soldiers and systems with smart technology across all five domains (land, sea, air, space, and cyber), to shorten decision cycles, better visualize the battlespace, and set the basis for a kill web.

Kill Web. A kill web is an AI-enabled kill chain that connects sensors and shooters to automatically execute targeting at machine speed. An AI-enabled kill web will transform warfare. The AI will synchronize the effects of many networked munitions in time, space, and purpose and will speed up sensor-to-shooter timing exponentially.

Masking. The full spectrum, multidomain effort to **deceive** enemy sensors and **disrupt** enemy targeting. Masking requires commanders to take action to deceive and disrupt. Masking is essential to survive and win in the modern battlespace and should be a principle of war in the 21st century. (Author's definition)

Mesh CP Configuration. The distribution of the CP infrastructure into resilient "functional nodes" that are small, armored, mobile, dispersed, and masked throughout the battlespace, yet remain in effective communication, with each other in the execution of mission command and are ready to assume command when required. Each node becomes a means to mitigate the risk of the nullification of other nodes. The ideal mesh CP configuration is a flexible self-forming, self-healing, and eventually self-organizing tactical network arrangement of command nodes. (Author's definition)

Mission Command. There are several definitions of Mission Command:

> **US Joint Publication (JP) 0-1:** "Mission Command is the conduct of military operations through decentralized execution based upon mission-type orders. Mission Command exploits the human element … emphasizing trust, force of will, initiative, judgment, and creativity. Successful Mission Command demands that subordinate leaders at all echelons exercise disciplined initiative and act aggressively and independently to accomplish the mission. They focus their orders on the purpose of the operation rather than on the details of how to perform assigned tasks. Essential to Mission Command is the thorough understanding of the commander's intent at every level of command and a command climate of mutual trust and understanding." (JP 1-0)
>
> **US Joint Staff J7:** "Mission Command provides the means through commander's intent, mission type orders, and decentralized execution to operate at the speed of the problem." From US Joint Staff J7, *Second Edition of the Insights and Best Practices Focus Paper on "Mission Command"* J7, Deployable Training Division (DTD) of the Joint Staff J7, Second Edition, January 2020.
>
> **US Army:** "Mission Command is the Army's approach to command and control that empowers subordinate decision making and decentralized execution appropriate to the situation." Army Doctrine Publication No. 6-0, *Mission Command: Command and Control of Army Forces*, Headquarters Department of the Army Washington, D.C., July 31, 2019, 1–3.
>
> **US Army Press:** "First and foremost, Mission Command is a leadership philosophy. It is a mindset for leading a team in a way that facilitates followers to exercise initiative within their leader's intent. More specifically, it requires leaders to provide a vision of what they ultimately want to accomplish along with a minimum level of instruction that dictates the how. To be successful, this requires not only leadership, but followership as well. Subordinates must remain disciplined to achieve the leader's vision, which includes adhering to Army Values and standards and knowing when to seize unforeseen opportunities or react to unanticipated threats … The principles of Mission Command are: build cohesive teams through mutual trust; create shared understanding;

provide a clear commander's intent; exercise disciplined initiative; use mission orders; and accept prudent risk. Leaders cannot just execute these principles at that crucial time when the need is greatest, they need to already have a solid foundation built within the organization … Mission orders are directives that emphasize results to be obtained, not how to achieve them. They explain how the leader wants to achieve a decision in time and space—everything else is secondary." Nathan K. Finney and Jonathan P. Klug, *Mission Command in the 21st century: Empowering to Win in a Complex World*, US Army, The Army Press, Fort Leavenworth, KS, 2016, vii, 2.

Author's Definition: Mission Command is a philosophy of the art of command that decentralizes authority and responsibility to trained and competent leaders and followers to accomplish the mission with minimum direct control from the higher commander.

Some notes about Mission Command from the author:

1. Mission Command requires extensive practice and training of the commander, subordinate leaders, and team members.
2. Leaders at every echelon must understand the mission, intent, and operational concept one and two levels higher. This understanding makes it possible to exercise disciplined initiative.
3. A visualization of the dynamics of Mission Command (shown below) can aid commanders in understanding the relationship of the level of training, command wisdom (experience and education), disciplined initiative (initiative guided by the commander's intent), and intent (two levels higher), required to execute successful Mission Command.
4. Based upon this visualization, Mission Command is the execution of loose control based on intent-based planning, trust, education, and the development of initiative and expertise.
5. The art of command is knowing how, where, and when to move from tight control to loose control, to make decisions in time, and to observe, orient, decide, and act faster than the enemy to accomplish a mission.
6. Mission Command is dynamic. Effective command adjusts styles from tight control to loose control based upon the maturity, skill, and education of subordinate leaders and the training level of the unit.
7. The acid test of Mission Command is that a unit can still accomplish the mission, when led by a corporal, if the officers and sergeants are dead or disabled.

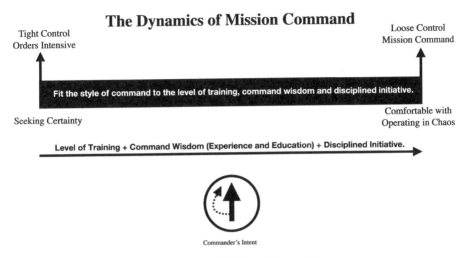

Dynamics of Mission Command

Mobile Striking Power. The offensive capacity of any military system, unit, or force to generate offensive action to move across the battlespace and disable or destroy the enemy.

Multidomain Operations. Multidomain Operations (MDO) is the combined-arms employment of joint and Army capabilities to create and exploit relative advantages that achieve objectives, defeat enemy forces, and consolidate gains on behalf of joint force commanders. (US Army Field Manual 3-0 Operations dated October 1, 2022).

Striking Power. The offensive capacity of any military system, unit, or force to generate offensive action to disable or destroy the enemy. Striking power is essential to offensive action. Although defense may be the strongest form of war, we do not win wars by defending. Only offensive action brings victory. (Author's definition)

Super Swarm. A super swarm is a group of AI-enabled, disposable, autonomous robotic systems that cooperate and are directed by a single "pilot" to attack multiple targets simultaneously from different angles. The robotic systems in the swarm act as "intelligent agents" performing actions to achieve goals, which are set by decisions made by the AI. A super swarm combines the tactic of swarming (convergent, multiple, and relentless attack) in a network of intelligent agents. To accomplish this, AI organizes, navigates, synchronizes, and directs the super swarm. One human "pilot," or the AI, steers the swarm and regulates its activation or deactivation.

Tactical Operations Center (TOC). A tactical command post.

Colonel John Antal, US Army (Ret.) briefs armor and cavalry leaders on the changing methods of warfare at the US Army's Maneuver Center of Excellence at Fort Moore, Georgia, on March 14, 2023. To contact the author, go to johnantal.com.

Recommended Resources

Videos

Lessons of the Second Nagorno-Karabakh War, US Army Fires Conference, accessed September 2021, at https://lnkd.in/gmQGvmeW.

John Antal, *Azerbaijan and Armenia*, Maneuver Warfighter Conference, accessed February 15, 2022, at https://youtu.be/_At9txsUKIw; and https://lnkd.in/gQVcfFEt.

John Antal presentation at the Maneuver Warfighter Conference (MWFC), accessed September 14, 2022, at https://youtu.be/EOrc4x-DlmI.

Podcasts

Top Attack: Lessons Learned from the Second Nagorno-Karabakh War, Army Futures Command Mad Scientist Podcast 2021 with Col. John Antal (Ret.), accessed April 1, 2021, at https://lnkd.in/eHEFqmr.

Sooner Than We Think: Command Post Survivability and Future Threats with COL (Ret.) John Antal, Army Futures Command Mad Scientist Podcast 2022, accessed May 4, 2022, at https://madsciblog.tradoc.army.mil/410-sooner-than-we-think-command-post-survivability-and-future-threats/.

Books

Antal, John. *7 Seconds to Die: A Military Analysis of the Second Nagorno-Karabakh War and the Future of Warfighting.* Oxford: Casemate Publishers, 2022.

Selected Bibliography

Ahronheim, Anna. "Israel's Operation Against Hamas was the World's First AI War." *The Jerusalem Post*, May 26, 2021.
Alderman, Ray. "Transitioning from the Kill Chain to the Kill Web." *Military Embedded Systems*, May 30, 2018.
Alexander, Bevin. *How Great Generals Win*. New York: Avon Books, 1993.
Allen, Gregory C. "Understanding China's AI Strategy, Clues to Chinese Strategic Thinking on Artificial Intelligence and National Security." *Center for New American Security*, February 2019.
"Analysis Russian Afganit active protection system is able to intercept uranium tank ammunition TASS 11012163." *Army Recognition*, December 10, 2016.
Anderson, Chris. "Ray Kurzweil on What the Future Holds Next." TED2018 Interview, TED Conferences, LLC, 2018.
Antal, John. *7 Seconds to Die: A Military Analysis of the Second Nagorno-Karabakh War and the Future of Warfighting*. Oxford: Casemate Publishers, 2022.
Antal, John. *7 Leadership Lessons of the American Revolution*. London: Casemate Publishers, 2013.
Antal, John. *Leadership Rising: Raise Your Awareness, Raise Your Leadership, Raise Your Life*. Oxford: Casemate Publishers, 2021.
Army Science Board. *Multi Domain Operations, Final Report*. Department of the Army, Office of the Deputy Under Secretary of the Army, Washington, D.C., May 2019.
Arrol, Matthew R. (LTC US Army), Milford Beagle (LTG US Army), Jason Slider (BG US Army). "The Graveyard of Command Posts What Chornobaivka Should Teach Us about Command and Control in Large-Scale Combat Operations." *Military Review Online*, March 2023.
Arquilla, John, and David Ronfeldt. *Swarming and the Future of Conflict*. Santa Monica: RAND Corporation, 2000.
Axe, David. "To Hide From Ukraine's Drones, Russian Troops Could Lay Smoke Screens." *Forbes*, December 17, 2021.
Balestrieri, Steve. "Remembering Adm. John Thach, Naval Aviator, Died on This Day 1981." *SOFREP*, April 15, 2019.
Ball, Geoff, Chad Skaggs, and USMC authors et al. "Signature Management (SIGMAN) Camouflage SOP: A Guide to Reduce Physical Signature Under UAS." Twenty-Nine Palms: The Warfighting Society, August 2020.
"Battlefield Modernization – Suite for Future Armored Vehicles." *Rafael Advanced Defense Systems Ltd.*, Haifa, Israel, accessed January 25, 2023.
Benitez, Mike. "OA-X: More Than Just Light Attack," *War on the Rocks*, August 16, 2016.
Beaumont, Peter. "Defenders of Mariupol are the Heroes of Our Time: the Battle that Gripped the World." *The Guardian*, May 17, 2022.
Boyd, Iain. "Chinese spy balloon over the US: An Aerospace Expert Explains How the Balloons Work and What They Can See." *The Conversation*, February 4, 2023.
Boyer, Peter J. "A Different War: Is the Army Becoming Irrelevant?" *The New Yorker*, June 23, 2002.

Brose, Christian. *The Kill Chain: Defending America in the Future of High-Tech Warfare.* New York: Hachette Books, 2020.

Brose, Christian. "Testimony of Christian Brose to the House Armed Services Committee Subcommittee on Cyber, Information Technologies, and Innovation the Future of War: Is the Pentagon Prepared to Deter and Defeat America's Adversaries." Washington, D.C.: US House of Representatives, Armed Services Committee, February 9, 2023.

Brown, Mike. "LIMPID Armor Creates HoloLens Helmet for Ukraine's Military." *Inverse*, November 3, 2016.

Brynjolfsson, Erik. "The Second Machine Age: Work, Progress, and Prosperity in a Time of Brilliant Technologies." LinkedIn Speaker Series talk, February 22, 2018.

BulgarianMilitary.com, "8 Russian 122mm Howitzers Destroyed by Ukrainian Drone Strikes," Bulgarian Government Publication, March 25, 2022.

Butusov, Yury. "The Russian Army killed 37 Ukrainian Soldiers near Zelenopillia." *Censor.NET*, Ukrainian news portal, July 11, 2014.

Calloway, Audra. "Picatinny 'Speed Bag' Resupplies Soldiers with Less Equipment Damage." US Army website, June 25, 2014.

Cancian, Mark F., and Eric Heginbotham. *The First Battle of the Next War: Wargaming a Chinese Invasion of Taiwan.* Center for Strategic and International Studies (CSIS), January 9, 2023.

"ChatGPT & Dall-e-2: Everything You Need to Know About the Newest Passion." *Techmango*, December 21, 2022.

Chen, Stephen. "Chinese Team Says Quantum Physics Project Moves Radar Closer to Detecting Stealth Aircraft." *South China Morning Post*, September 3, 2021.

Chu, Jennifer. "MIT engineers configure RFID tags to work as sensors; Platform may enable continuous, low-cost, reliable devices that detect chemicals in the environment." *MIT News*, Massachusetts Institute of Technology, June 14, 2018.

Clarke, M. L. *The Aeneid of Virgil, translated by John Dryden,* edited with introduction and notes by Robert Fitzgerald. London: Collier-Macmillan, 1968.

Cleary, Thomas trans. *The Art of War, Complete Texts and Commentaries, Sun Tzu.* Boston: Shambhala Publications Inc., 2003.

Collins, Col. Liam. "Russia gives Lessons in Electronic Warfare." *Association of the US Army*, July 26, 2018.

Congressional Research Service. "The Army's Optionally Manned Fighting Vehicle (OMFV) Program: Background and Issues for Congress." February 22, 2019.

Cook, Ellie. "How Russia's 'Marker' Combat Robots Could Impact Ukraine War." *Newsweek*, January 18, 2023.

Coram, Robert. *Boyd: The Fighter Pilot Who Changed the Art of War.* New York: Back Bay Books, 2004.

Cott, Hugh B. *Adaptive Coloration in Animals.* Methuen: Oxford University Press, 1940.

Cowhig, David. "PRC Defense: Starlink Countermeasures." David Cowhig's Translation Blog, May 25, 2022.

Cranny-Evans, Sam, and Dr. Thomas Withington. "Russian Comms in Ukraine: A World of Hertz." *Russia Report*, March 9, 2022.

Dangwal, Ashish. "China Sends Suspected Military Reconnaissance Balloons Over Taiwan Amid Russian Ops In Ukraine; Beijing Responds." *The EurAsian Times*, March 2, 2022.

Dominguez, Felipe. "Raytheon Intelligence & Space to build mobile 50kW-class laser for US Army." *Raytheon News*, September 7, 2021.

Doyle, John M. "General: Precise Sensors to Close Kill Chain is a Key Takeaway from Ukraine War." *Seapower, The Official Publication of the Navy League of the United States*, May 5, 2022.

Edwards, Sean J. A. *Swarming and the Future of Warfare.* Santa Monica: RAND Corporation, 2005.

Eversden, Andrew. "Inside the Army's Distributed Mission Command Experiments In, and Over, the Pacific." *Breaking Defense*, March 23, 2022.

Ferran, Lee. "Army's Ultra-Endurance Zephyr Drone Comes Down After 'Unexpected Termination' over Arizona Desert." *Breaking Defense*, August 23, 2022.

Fitzgerald, Paula M. "Range 37 at Fort Bragg tests skills of even the most experienced Green Beret." *army.mil*, February 19, 2010.

Freedberg, Sydney, J. "War Without Fear: DepSecDef Work on How AI Changes Conflict." *Breaking Defense*, May 31, 2017.

Freedberg, Sydney J. "Target Gone In 20 Seconds: Army Sensor-Shooter Test." *Speaking Defense*, September 10, 2020.

Freedman, David H. "US Is Only Nation with Ethical Standards for AI Weapons. Should We Be Afraid?" *Newsweek*, September 15, 2021.

Fuller, J. F. C. *Generalship: Its Diseases and Their Cure: A Study of The Personal Factor in Command*. Harrisburg: Military Service Publishing Company, 1936.

Episkopos, Mark. "Russia Is Ready to Receive Shiny New T-14 Armata Tanks." *The National Interest*, July 7, 2021.

Gardner, Elliot. "All clear: the onset of see-through armour." *Army Technology*, August 28. 2018.

Genys, Andrius. "ARENA Active Protection System." *Military-Today.com*, 2023.

Gonzales, Matt. "Mobile Satellite System Reduces Communications Gaps, Increases Naval Interoperability." US Marine Corps Systems Command, December 15, 2021.

Grau, Lester W., and Charles K. Bartles. "The Russian Reconnaissance Fire Complex Comes of Age." The Changing Character of War Centre, Pembroke College, University of Oxford, With Axel and Margaret Johnson Foundation, May 2018.

Gross, Judah Ari. "IDF Intelligence Hails Tactical Win in Gaza, Can't Say How Long Calm Will Last." *The Times of Israel*, May 27, 2021.

Hambling, David. "The next era of drones will be defined by 'swarms'." *BBC, Future Now*, April 16. 2017.

Hannas, William et al. "China's Advanced AI Research Monitoring China's Paths to 'General' Artificial Intelligence." *Center for Security and Emerging Technology*, July 2022.

"HAROP Loitering Munition System." Israel Aerospace Industries website, accessed January 23, 2023.

Harris, Peter. trans. *Sun Tzu: The Art of War*. London: Everyman's Library, 2018.

Harris, Marc et al. "Megacities and the United States Army: Preparing for a Complex and Uncertain Future." Chief of Staff of the Army Strategic Studies Group, June 2014.

Harrison, Dontavian. "AUSA 2021: Secretary of the Army's Keynote Speech 11 October 2021 (transcript)." *army.mil*, October 14, 2021.

Heinrich, Jacqui, and Adam Sabes. "Gen. Milley Says Kyiv Could Fall Within 72 hours if Russia Decides to Invade Ukraine." *Fox News*, February 5, 2022.

Hennigan, W. J. "Downed Chinese Balloon Part of Global Spy Operation, Pentagon Alleges." *Time*, February 8, 2023.

Hensholdt. "SETAS – See Through Armour System." Germany: Hensholdt Taufkirchen, January 25, 2023.

Hoehn, John R. "Joint All Domain Command and Control (JADC2), Washington D.C.: Congressional Research Service Report (CRS), July 1, 2021." *Congress.gov*, March 18, 2021.

Hoffman, Bryce G. *Red Teaming: How Your Business Can Conquer the Competition by Challenging Everything*. New York: Crown Business, 2017.

Hofstetter, Jeremy, and Adam Wojciechowski. "Electromagnetic Spectrum Survivability in Large-Scale Combat Operations." Fort Benning: Infantry, Winter 2020–2021.

Howard, Michael, and Peter Paret, trans. *On War, Carl von Clausewitz*. Princeton: Princeton University Press, 1989.

Hughes, Wayne P. Jr. *Fleet Tactics and Coastal Combat* (second edition). Annapolis: Naval Institute Press, 2000.

Insider Business, "Watch the Navy's LOCUST (Low-Cost UAV Swarming Technology), Launcher Fire a Swarm of Drones." *YouTube*, April 20, 2017.

IRONVISION 'See-Through' Head-Mounted Display Technology for 360° Situational Awareness and Advanced Operation of Armored Fighting Vehicles." Elbit Systems Ltd., Haifa, Israel, accessed January 25, 2023.

Johns Hopkins University. "COVID dashboard." At https://coronavirus.jhu.edu/map.html.

Johnson, David. "The Tank is Dead: Long Live the Javelin, the Switchblade, the …?" *War on the Rocks*, April 18, 2022.

Johnson, John. "Analysis of Image Forming Systems." Ft. Belvoir, in Image Intensifier Symposium, AD 220160 Warfare Electrical Engineering Department, US Army Research and Development Laboratories, October 6–7, 1958.

Joint Air Power Competence Center (JAPCC). "All Domain Operations in a Combined Environment." North Atlantic Treaty Organization JAPCC, September 2021.

Kahn, Lauren. "How Ukraine Is Using Drones Against Russia." *Council of Foreign Relations (CFR)*, March 2, 2022.

Kalo, Avi. *AI-Enhanced Military Intelligence Warfare Precedent: Lessons from IDF's Operation Guardian of the Walls*. San Antonio: Frost and Sullivan, 2021.

Kan, Michael. "Starlink on a Drone? This Company Is Working on the Idea." *PCMag.com*, November 2, 2022.

Karber, Phillip. "Lessons Learned From the Russo–Ukraine War." Historical Lessons Learned Workshop by Johns Hopkins Applied Physics Laboratory and the US Army Capabilities Center (ARCIC), July 6, 2015.

Karber, Phillip, and David Axe. "The Ukrainian Army Learned the Hard Way—Don't Idle Your Tanks When the Russians Are Nearby." *Forbes Magazine*, August 5, 2020.

Karlin, Mara. "The Kill Chain: An Interview with Christian Brose." Johns Hopkins University, School of Advanced International Studies, May 26, 2020.

Kasapoglu, Can. "Dangerous Drone for All Seasons: Assessing the Ukrainian Military's Use of the Bayraktar TB2." *Eurasia Daily Monitor*, Volume 19, Issue 36, March 16, 2022.

Keller, Jared. "The Army is Testing a Robot Mini-Tank Straight Out of 'Fast and the Furious.' The Ripsaw is Here and Ready to Rock." *Task and Purpose*, July 27, 2021.

Kendra Cherry, "Nomophobia: The Fear of Being Without Your Phone." *Verywell Mind*, February 25, 2020.

Khaled, Fatima. "Ukraine Official Asks Elon Musk for Starlink Stations Amid Russian Invasion." *Newsweek*, February 26, 2022.

Khandelwal, Swati. "How A Drone Can Infiltrate Your Network by Hovering Outside the Building." *The Hacker News*, October 7, 2015.

Konaev, Margarita, and Kirstin J. H. Brathwaite. "Russia's Urban Warfare Predictably Struggles Fighting in Cities is Hard for any Military." *Forbes Magazine*, April 4, 2022.

Krause, Merrick. "Decision Dominance: Exploiting Transformational Asymmetries." *Defense Horizons*, The Center for Technology and National Security Policy National Defense University, February 2003.

Laird, Robin, and Edward Timberlake. *A Maritime Kill Web Force in the Making: Deterrence and Warfighting in the XXIst Century*. Pennsauken: Bookbaby, 2022.

"Lanius, Drone-based Loitering Munition for Complex Environments," *Elbit Systems*, accessed March 2, 2023.

LaPorta, James, et al. "Military Units Track Guns Using Tech that Could Aid Foes." *The Associated Press*, September 30, 2021.

"Legion-X, Multidomain Autonomous Network Combat Solutions for Unmanned Heterogeneous Swarms." *Elbit Systems*, accessed March 2, 2023.

Liang, Qiao and Wang Xiangsui. *Unrestricted Warfare*. Beijing: PLA Literature and Arts Publishing House, 1999.

Lumbard, Tanya. "I Corps Tests Distributed Mission Command concept in Indo-Pacific." *US Army website*, March 2, 2022.

Lewis, Jeffrey et al. "China's Growing Missile Arsenal and the Risk of a 'Taiwan Missile Crisis.'" *Nuclear Threat Initiative Report*, November 18, 2020.

Litovkin, Dmitry. "Infrared and Invisibility: Russia's New Tanks Top Up on Technology." *Russia Beyond*, May 17, 2017.

Losey, Stephen. "US Special Operations Command chooses L3Harris' Sky Warden for Armed Overwatch Effort." *Defense News*, August 1, 2022.

"M5 Ripsaw, Extreme Mobility, Decisive Lethality, Wingman Ready," *Team Ripsaw*, Textron Systems, accessed March 2, 2023.

M-19 Abrams Reactive Armor Tile (ARAT)." Ensign-Bickford Aerospace & Defense Company, accessed January 25, 2023.

"MAARS Weaponized Robot." *QinetiQ Group plc*, Cody Technology Park, Farnborough, Hampshire, United Kingdom, accessed January 25, 2023.

MacArthur, Douglas. "MacArthur Endorses More Aid to Britain; General Cables Warning Against 'Fatal Epitaph, Too Late.'" *The New York Times*, September 16, 1940.

Mack, Erick. "US Military Says SpaceX Handily Fought Off Russian Starlink Jamming Attempts." *CNET*, April 22, 2022.

Malyasov, Dylan. "Ukrainian Bayraktar TB2 Drones Successfully Attack Russian Convoys." *Defense Blog*, February 28, 2022.

Mamontov, Sergey. "War machine: Robots to Replace Soldiers in Future, says Russian Military's Tech Chief." *RT News*, July 6, 2016.

McCardle, Guy. "Russian Explosive Reactive Armor Explained." *SOFREP*, September 16, 2022.

McKenzie, Gen. Kenneth F., Jr. "Posture Statement of US Central Command before the Senate Armed Services Committee." April 22, 2021.

McRobbie, Linda Rodriguez. "When the British Wanted to Camouflage Their Warships, They Made Them Dazzle" *Smithsonian Magazine*, April 7, 2016.

"MetaTrader 5 Trading Platform." *metaquotes.net*, January 23, 2023.

Milley, Mark A. "AUSA Eisenhower Luncheon, 4 October 2016." Washington, D.C.: Association of the United States Army, October 4, 2016.

Mitzer, Stijn, and Joost Oliemans. "Defending Ukraine – Listing Russian Military Equipment Destroyed By Bayraktar TB2s." *Oryx website*, February 27, 2022.

"Multi-Utility Tactical Transport (MUTT)." General Dynamics Land Systems, Sterling Heights, Michigan, accessed January 25, 2023.

Munoz, Carlo. "US Navy, CENTCOM Seek Solutions for Stratospheric ISR Operations." *Janes*, April 5, 2021.

Muwashi. "UPRISE Passive Load-Bearing Exoskeleton." Quebec, Canada, accessed January 25, 2023.

Neikirk, Todd. "The Germans Had High Hopes for the Goliath Tracked Mine." *War History Online*, June 6, 2022.

Ochsner, Evan. "DOD to prioritize non-kinetic C-UAS options in 2022." *Inside Defense*, January 12, 2022.

"OFFensive Swarm-Enabled Tactics (OFFSET)." Defense Advanced Research Projects Agency (DARPA), accessed March 2, 2023.

Oh, Joshua. "Distributed C2 Concept to Address the Pacific Theater." *US Army*, October 17, 2022.

Osborn, Kris. "Army Evaluates Newly Unveiled AbramsX Main Battle Tank for Future War Into 2050, General Dynamics Land Systems Unveils Breakthrough Abrams for Future." *Warrior Maven*, October 11, 2022.

Parkinson, Malcolm. "The Artist at War." *Prologue Magazine* in The National Archives, Vol. 44, No. 1, Spring 2012.

Parsons, Dan. "What Ukraine Is Teaching US Army Generals About Future Combat: US soldiers can expect to be under constant enemy surveillance and threatened by long-range precision artillery in the next war." *The War Zone*, October 14, 2022.

Pellerin, Cheryl. Third Offset Strategy Bolsters America's Military Deterrence." *US Department of Defense (DOD) News*, October 31, 2016.

Podolyak, Mykhailo. "If something is launched into other countries' airspace, sooner or later unknown flying objects will return to (their) departure point." @Podolyak_M, Twitter. December 5, 2022.

Ranjan, Rohit. "Russia Claims 2,911 Ukrainian Military Infrastructure Facilities Destroyed." *Republicworld.com*, March 10, 2022.

Roblin, Sébastien. "In 1944, USS *Intrepid* Survived Assault from a Huge Japanese Warship." *The National Interest*, June 6, 2021.

Roblin, Sébastien. "Israel's Bombardment Of Gaza: Methods, Weapons And Impact." *Forbes*, May 26, 2021.

"ROSY—Rapid Obscuring System Survive On The Move." *Rheinmetall Waffe Munition GmbH.*, 2021.

Rubin, Uzi. "The Second Nagorno-Karabakh War: A Milestone in Military Affairs." *Mideast Security and Policy Studies No. 184*, The Begin-Sadat Center for Strategic Studies, December 2020.

Saballa, Joe. "US Air Force Tactical Fighter Flown by Artificial Intelligence for First Time." *The Defense Post*, February 14, 2023.

Sanders, Patrick. "Chief of the General Staff Speech at RUSI Land Warfare Conference." *Gov.UK*, June 28, 2022.

Saran, Cliff. "Stanford University's *AI Index 2019* annual report has found that the speed of artificial intelligence (AI) is outpacing Moore's Law." *Computerweekly.com*, December 12, 2019.

Satter, Raphael and Dmytro Vlasov. "Ukraine Soldiers Bombarded by 'Pinpoint Propaganda' Texts." *AP News*, May 11, 2017.

Scharre, Paul. "Centaur Warfighting, The False Choice of Humans vs. Automation." *Temple International and Company L.J.*, 2016.

Scharre, Paul. "Robotics on the Battlefield Part II: The Coming Swarm." Center for a New American Security (CNAS), October 15, 2014.

Scott, Benjamin. Army Counter-UAS, 2021–2028." *Military Review*, March–April 2021.

"Sentinels of the Sky: The Persistent Threat Detection System." *Lockheed Martin website*, accessed March 2, 2023.

Sevryukov, Lt. Gen. Sergei. "Statement by First Deputy Chief of Main Military-Political Directorate of Russian Armed Forces Lieutenant General Sergei Sevryukov." *mil.ru*, Russian Ministry of Defense Video 4, January 2022.

Shoaib, Alia. "An Elite Ukrainian Drone Unit Exploits the Cover of Night to Destroy Russian Tanks and Trucks While Their Soldiers Sleep." *Business Insider*, March 20, 2022.

Shoaib, Alia. "Inside the elite Ukrainian Drone Unit Founded by Volunteer IT Experts: 'We are all soldiers now.'" *Business Insider*, April 9, 2022.

SHTORA-1 Active Defense System," *Defense-Update.com*, October 12, 2005.

Singh, Mandip. "Global MALE & HALE UAV Key Developments Across Global Top 10 Defense Spenders." *European Security and Defense*, January 10, 2023.

Simonite, Tom. "How Starlink Scrambled to Keep Ukraine Online: Elon Musk's intervention demonstrates how satellite internet could route around war or censorship far beyond Ukraine." *Wired*, May 11, 2022.

Smith, Rupert. *The Utility of Force: the Art of War in the Modern World*. New York: Knopf Doubleday Publishing Group, 2008.

Sorley, Lewis. *Press On! Selected Works of General Donn A. Starry, Volume I*. Fort Leavenworth: US Army Combined Arms Center, Combat Studies Institute, 2009.

Spencer, David K., Stephen Duncan, Stephen, and Adam Taliaferro. "Operationalizing Artificial Intelligence for Multidomain operations: a First Look." Proc. SPIE 11006, Artificial Intelligence and Machine Learning for Multidomain Operations Applications, May 10, 2019.

Spires, John. "DJI Named Top Commercial Drone Maker with 70% Market Share." *DroneDJ*, October 16, 2020.

Stace, Leanne. "BATTLEVIEW 360." *BAE Systems*, BAE Systems, Inc. US, Falls Church, Virginia, accessed January 25, 2023.

Stepanenko, Kateryna et al. "Russian Offensive Campaign Assessment, September 30, 2022." Institute for the Study of War (ISW), September 30, 2022.

Sutton, H.I., "Why Ukraine's Remarkable Attack On Sevastopol Will Go Down In History." Publication/Org?, November 17, 2022.

Tegler, Eric. An Army General Says The Robotic Combat Vehicles It's Experimenting With Will Be The 'Ghosts Of Patton's Army.'" *Forbes Magazine*, July 30, 2021.

TERRA RAVEN Multi-Function Counter-Measures (MFCM)," *BAE Systems*, Inc. US, Falls Church, VA, accessed January 25, 2023.

Tingley, Brett. "Jet-Powered Coyote Drone Defeats Swarm In Army Tests" *The War Zone*, thedrive.com, July 26, 2021.

TRADOC G2. *The Red Team Handbook, The Army's Guide to Making Better Decisions*. US Army Training and Doctrine Command, Version 9, 2019.

Trevithick, Joseph. "China Conducts Test Of Massive Suicide Drone Swarm Launched From A Box On A Truck." *The War Zone*, thedrive.com, October 14, 2020.

Trevithick, Joseph. "Massive Drone Swarm Over Strait Decisive In Taiwan Conflict Wargames: Air Force and Independent Think Tank Simulations Show Giant Drone Swarms are Key to Defeating China's invasion of Taiwan." *The War Zone*, thedrive.com, May 19, 2022.

Tsouras, Peter G. *Warriors' Words, A quotation Book, From Sesostris to Schwarzkopf, 187 BC to AD 1991*. London: Arms and Armour Press, 1992.

Tsukanov, Ilya. "Chinese Military 'Deeply Alarmed Over Musk's Starlink Satellites Dual-Use Capabilities." *Sputnik International*, May 11, 2022.

Turek, Matt. "Explainable Artificial Intelligence (XAI)." Defense Advanced Research Projects Agency (DARPA), accessed March 2, 2023.

Ukraine says it hit Russian Command Post." *Associated Press News*, April 23, 2022.

US Pacific Fleet. Accessed February 13, 2023, at https://www.cpf.navy.mil/About-Us/.

US Army, *Field Manual 3-0, Operations*. Headquarters Department of the Army, Washington, D.C. October 1, 2022.

US Army. *Doctrine Publication No. 6-0, Mission Command: Command and Control of Army Forces*. Washington, D.C.: Headquarters Department of the Army, 2019.

US Army Training and Doctrine Command. *ATP 7-100.3 Chinese Tactics*. Washington, D.C.: US Army, 2021.

Venckunas, Valius. "Baykar drone factory in Ukraine to be complete in two years: CEO." *Aerotime Hub*, October 28, 2022.

Viereck, George Sylvester. "What life Means to Einstein, an Interview by George Sylvester Viereck." *The Saturday Evening Post*, 1929.

Wall, Brandon and Nicholas Ayrton. "Drones and Starlink: Combining Satellite Constellations with Unmanned Navy Ships." Center for International Maritime Security, September 1, 2021.

Wohlstetter, Roberta Morgan. *Pearl Harbor, Warning and Decision*. Stanford: Stanford University Press, 1962.

Yang, William. "China: Fears Grow for Detained Anti-COVID Protesters." *Deutsche Welle*, January 17, 2023.

Yilmaz, Harun. "No, Russia will not invade Ukraine: A large-scale military operation does not fit into Moscow's cost-benefit calculus." *Al Jazeera*, February 9, 2022.

Index

AbramsX, 142–43
acoustic spectrum, 27
ADCOP (All-Domain Common Operational Picture), 99, 100–4, *105*, 192
    and command posts, 177, 191
Admiral Makarov (frigate), 85–87
ADO (All Domain Operations), 70–73
Aerorozvidka (aerial reconnaissance), 146
Afghanistan, 7, 116, 198
AFVs (armored fighting vehicles), 141–42
AGI (Artificial General Intelligence), 61
AI (artificial intelligence), 8, 61–65, 67–68
    and decision dominance, 115, 116
    and Israel–Hamas War, 56–60
    and kill webs, 79–83
    and leadership, 190
    and MDO, 70–73
    and RCVs, 166–67
    and speed, 104–5
    and super swarms, 92–94
    and urban warfare, 145
aircraft, 153–59
Alderman, Ray, 80
Alexander the Great, 188
American Indian Wars, 88–89
American Revolutionary War, 95–97, *98*, 99
Angel Has Fallen (film), 91–92
ANI (Artificial Narrow Intelligence), 61–62
animal kingdom, 18
Arestovych, Oleksiy, 77
Arizona, USS, 5

Arma 3, 120–21
Armenia, 7–8, 25, 38, 114, 200
ASI (Artificial Super Intelligence), 61
ASPI (Australian Strategic Policy Institute), 65
ATGMs (anti-tank guided missiles), 161
atomic bomb, 64
autonomous weapons, *64*, 67–73
Ayrton, Nicholas, 133–34
Azerbaijan, 7–8, 25, 78, 200
    and decision dominance, 112, 114
    and first strike, 37–38, 43
    and UAVs, 49

balloons *see* stratospheric balloons
Baltic states, 35
battleshock, 10, 11
Belarus, 35
Benitez, Maj Mike, 155
Big Blue Blanket, 154, 155, 158, 191
binoculars, 18
Black Sea Fleet, 77, 85–87
BMI (brain–machine interface), 100
British Army, 122
Brose, Christian:
    Kill Chain: Defending America in the Future of High-Tech Warfare, 63, 77
Brunson, Lt. Gen. Xavier, 107
Brynjolfsson, Erik, 106

Caesar, Julius, 188
camouflage, 18–21, 22

cell phones *see* mobile phones
centaur approach, 106–7, 108
ChatGPT (Chat Generative Pre-trained Transformer), 62–63, 68
China, 9, 34
    and AI, 62–63, 64–65
    and drones, 146
    and kill webs, 80
    and leadership, 188, 189
    and missile range, *37*
    and quantum spectrum, 27, 29
    and stratospheric balloons, 149
    and Taiwan, 44, 199
    see also People's Liberation Army (PLA)
cities *see* urban warfare
Coffman, Maj. Gen. Ross, 163
Cold War, 64
Collins, Col. Liam, 16
commanders, 117–19
control, 103
Cornwallis, Gen. Charles, 97
Cott, Hugh, 19
coup d'oeil, 95–97, *98*, 99
COVID-19 pandemic, 101
Cowpens, Battle of, 95–97, *98*, 99
CPs (command posts), 75–77, 169–74, *175*, 176–78, 190–91
Crimea, 34, 35
CSIS (Center for Strategic and International Studies), 36
CTCs (combat training centers), 36–37, 193
CUAS (counter-unmanned aerial systems), 51, 52–54, 154–59, 178, 191
Custer, Lt Col. George Armstrong, 88–89
cyberattacks, 129–30

Dall-E, 68
DARPA (Defense Advanced Research Projects Agency), 145–46
data, 81–82, 102
DDoS (Distributed Denial of Service), 129–30

decision dominance, 111–17, 120–23
    and commanders, 117–19
decision making, 99–100
design, 191–92
deterrents, 33
disrupters, 11–12
disruptive camouflage, 19, 20
Distributed Mission Command, 170
DL (deep learning), 61–62, 67
double envelopment, 95–97, *98*, 99
drones, 13–14, 16, 41–42
    and CUAS, 154–57
    and field testing, *79*
    and swarms, *82*, 87–94
    and urban warfare, 146–47

Edwards, Sean J. A., 88
Einstein, Albert, 1
Elbit Systems Ltd, 143–44
electromagnetic energy, *22*, *23*, 25–26
electronic signature, 24–27
electronic warfare jamming, 52
Esper, Mark, 163
Estonia, 35

Fedorov, Mykhailo, 131, 133
feedback, 102–3
First Indochina War, 19
first strikes, 33–38, 42–45
forging battleshock, *168*, 194–98
    and leadership, 188–91
    and training, 192–94
    and weaponry design, 191–92
Formosa *see* Taiwan
France, 19, 164, 167
Fuller, J. F. C., 170, 190

gaming, 101–2
Gaza *see* Israel–Hamas War
Genda, Cmdr Minoru, 1, 2–4
Gerasimov, Gen. Valery, 77

Germany, 19, 33, 64
Great Britain, 19; *see also* British Army; Royal Navy
Great War *see* World War I
Grimshaw, Sgt Danielle, *40*

HAA (High-Altitude Airship), 150
Hamas, 8, 50, 56, 115
Heckl, Lt. Gen. Karsten, 77
Hill, Vice Adm Jon, 79–80
"HiLo" mix, 149–50
holistic view, 101
Honchar, Yaroslav, 131
HOTL (human-on-the-loop) robots, 163, 164
HVTs (high-value targets), 42–44, 56–57, 58
Hype Cycle, 68, 73

IADS (integrated air-defense system), 154
imagination, 1, 3, 7
Imperial Japanese Navy (IJN), 1–6
intelligence, 44–45
internet, 65, 115, 129–30
Intrepid, USS, 153–54
IOBT (Internet of Battlefield Things), 79, 80
Iran, 89–90, 146, 188, 189
Iraq, 7, 36–37, 164, 198
Israel, 114, 143–44, 146
Israel Defense Forces (IDF), 8, 50, 55–60
Israel–Hamas War, 8, 9, 12, 56–57, 194–95
    and decision dominance, 115
    and kill webs, 80–81
    and UCAVs, 50
    and urban warfare, 138–39
ISR (Intelligence, Surveillance and Reconnaissance), 9, 15–16, 27
    and the sky, 146–50
    and Taiwan, 125–28
    and urban warfare, 143–46, 150–52

JADC2 (Joint All-Domain Command and Control), 80, 104–9
Japan, 1–6, 21, 33

and aircraft, 153–54, *155*
and Malaya Campaign, 111–12, *113*

Kaliningrad, 35
kamikaze attacks, 153–54, *155*
Karber, Dr. Phillip A., 16
Khalitov, Vyacheslav, 23
kill chains, 77–9, 101, 143
kill webs, 67, 71–72, 79–83
kinetic technologies, 53
Kochavi, Lt. Gen. Aviv, 60, 81
Konashenkov, Igor, 43
Korean War, 34
Kurzweil, Ray, 68
Kuwait, 34

LANIUS, 144–45
lasers, 51–52
Latvia, 35
leadership, 188–91, 195
LegionX, 143–44
Lithuania, 35
Litovkin, Dmitry, 133
Live Synth, 119
LOCUST (Low-Cost UAV Swarming Technology), 92
loitering munitions, 40–41, 42, 49–50, 191
    and LANIUS, 144–45
LRPFs (long-range precision fires), 9, 67–68, 69, 70, 73
LTA (lighter-than-air) systems, 150

Maginot Line, 164, 167
Malaya, 111–12, *113*
masking, 10, 11, 29–30
    and command posts, 170–71, 173–74
    and fully autonomous weapons, 69–70, 71–73
MBT (main battle tank), 139–40, 142–43
McConville, Gen. James, 115, 169
MDO (Multidomain Operations), 68–71, 197–98
microwave, 52

Milley, Gen. Mark, 8, 170
missiles, 157–58
Mission Command, 170, 174, 176, 190, 192–94
ML (machine learning), 61–62, 67
mobile phones, 13–14, 16–17, 25, 65
mobile striking power, 10–11, 139–43
Moore's Law, 68
Morgan, Brig Gen. Daniel, 95–97, 99
Moskva (cruiser), 50–51
MT5 (Meta-Trader 5), 103
multidomain capacities, 8–9
Multiple Launch Rocket Systems (MLRS), *34*
MUM-T (Manned Unmanned Teaming), 163, 164–65
Musashi, Miyamoto, 3
Musk, Elon, 43, 62, 115, 128; *see also* Starlink

Nagorno-Karabakh *see* Second Nagorno-Karabakh War
Nagumo, Adm Chūichi, 4
Napoleon Bonaparte, 188
NATO, 29, 35, 156, 198–99
navigation, 14, 52, 102
Nelson, Adm Lord Horatio, 63
North Korea, 35, 188, 189, 199
Norwegian Army, 121, 122
NOTM (Networking on the Move), 116

obscuration, 176–77
OMFVs (optionally manned fighting vehicles), 161, 162–63, 166
operations: *Guardian of the Walls* (2021), 8, 55–60, 63
optical signature, 18–22

Pearl Harbor, 2–6, 44–45
People's Liberation Army (PLA), 90, 133, 143
  and Taiwan, 35–36, 125–28
Petrov, Maj Sergey Ivanovitch, 13–14
Philippine Sea, 153
Podolyak, Mykhailo, 21

Poland, 35
Putin, Vladimir, 30

quantum spectrum, 27, 29
Quest, 121–23

radio frequency, 24
Rainey, Gen. James, 20–21
RCVs (Robotic Combat Vehicles), *160*, 161–63, 164, 165–67
Reconnaissance Strike Complex, 35
red teaming, 173
Republic of Korea, 33–34, 199
Rift, 121–23
robotic systems, 7–8, 68, *93*
  and design, 191–92
  and tanks, 142–43
  and urban warfare, 143–44
  see also RCVs
Romania, 35
Royal Navy, 19, 112
Rubin, Uzi, 90
Russia, 35, 188
Russian Army: 49th CAA, 77
Russia–Ukraine War, 7, 8–9, 12, 194–95
  and command posts, 169, 170, 171, 174
  and CUAS, 157
  and cyberattacks, 129–30
  and decision dominance, 115
  and drones, 146
  and electronic signature, 25–26
  and first strike, 43–44
  and kill chains, 77
  and leadership, 189
  and masking, 29–30
  and NATO, 198–99
  and Russian Federation, 200
  and Sevastopol, 85–87, 90–91
  and Starlink, 130–31, *132*, 133, 134
  and surprise strikes, 34
  and thermal signature, 23–24
  and transparent battlespace, 13–18, 21–22
  and UAVs, 49

and UCAVs, 39–40, 50–51, 53–54
and urban warfare, 135–38, 139, 140, 143, 145, 151
SA (situational awareness), 140–41
Saddam Hussein, 34
Santa Fe, USS, *28*
Saudi Arabia, 89–90
Scharre, Paul, 107, 152
SDR (software-defined radio), 144
Second Nagorno-Karabakh War, 7–8, 9, 12, 194–95
   and CUAS, 53
   and decision dominance, 112, 114
   and electronic signature, 25
   and first strike, 37–38, 43
   and kill chains, 78
   and robotic forces, *93*
   and swarming, 90
   and UAVs, 49
   and UCAVs, 50
   and urban warfare, 138, 143
seismic spectrum, 27
sensor networks, 9, *17*, 18
   and command posts, 170–71
   and first strikes, 35, 36
September 11 attacks, 45
Sevastopol, 85–87, 90–91
Sevryukov, Lt. Gen. Sergei, 17
Shinsheki, Gen. Eric, 73
Shotwell, Gwynne, 131, 133
simulation, 118, 119, 120–23, 193–94
Singapore, 111–12
Sky Warden, *157*, 158
smoke, 176–77
SOCOM (Special Operations Command), 156, 158
South China Sea, 35
SpaceX, 130, 131, 134
speed, 104–5, 117
Sprague, Capt Tom, 153
Starlink, *128*, *129*, 130–31, *132*, 133–34
Starry, Gen. Donn A., 174

Starshield, 134
stratospheric balloons, 125–28, 146–50
striking power, 10–11
submarine warfare, 26, *28*
Sun Tzu, 73, 118, 139
super swarms, 91–94
support units, 195–97
Suvorov, Alexander, 104
Suwalki Corridor, 35
swarming, *82*, 87–94, 145–46
Syria, 164

Taiwan, 9, 35–36, 44, 199
   and ISR, 125–28
   and super swarms, 92
   and urban warfare, 151
   and World War II, 153–54
tanks, 23, *30*, 139–43
Tatum, Capt. Eric, *79*
telescopes, 18
tempo, 60–61
terrorism, 45
Thach, Cmdr "Jimmy," 154
thermal signature, 22–24
time, 63
TITAN (Tactical Intelligence Targeting Access Node), 80
Tnufa (Momentum), 60
top attack, 49–54
TOS multiple rocket launchers, *15*
training, 117–19, 120–23, 192–94
   and command posts, 172, 176
   and support units, 195–97
transmission security (TRANSEC) techniques, 26–27
transparent battlespace, 13–18, 21
tripwires, 33
Tsuji, Lt Col. Masanobu, 112
TTP (tactics, techniques, and procedures), 21, 178
tunnel networks, 57, 58
Turek, Dr. Matt, 145

Turkey, 7, 80, 114, 146, 200

UAS (unmanned aerial systems), 9, *10*, 49
  and AFVs, 142
  and command posts, 176
  and swarming, 90
UAVs (unmanned aerial vehicles), 38, *43*, 44, 49
  and IDF, 58–59
  and swarming, 89–90
UCAVs (unmanned combat aerial vehicle) systems, 18, 22, 38–40, 41–42, 49–51
UI (user interface), 100, 101–3
Ukraine *see* Russia–Ukraine War
Ukrainian Army, 14–15, 16
United States of America (USA), 1–6, 7, 65, 198–99
  and atomic bomb, 64
  and Chinese ISR, 128
  and Doolittle raid, 21
  and drones, 146
  and JADC2, 104–5, 107–8
  and kill webs, 80
  and Russia–Ukraine War, 8
  and Taiwan, 35–36
urban warfare, 8, 138–39
  and Bakhmut, 135–38
  and command posts, 176
  and ISR, 143–46, 150–52
  and the sky, 146–50
  and tanks, 140–41
US Air Force (USAF), 20, 133, 154–57
US Army:
  and aircraft, 156, 158
  and camouflage, 19–20
  and CTCs, 36–37
  and decision dominance, 114–15, 116–17
  and kill chains, 77–78
  and MDO, 70
  and MUM-T, 164–65
  and robots, 163
  and stratospheric ISR, 147–49
  and training, 193–94, 195–96
  and tripwires, 33
US Marine Corps, 20, *40*, *194*
US Navy, 2–6, 26
  and camouflage, 19, 20
  and Rift, 121–22
  and swarming, 92
  and Taiwan, 153–54
USVs (unmanned surface vessel), 87

VBS3 (Virtual Battlespace 3), 120
Vietnam War, 7, 19
Vimeur, Gen. Jean-Baptiste-Donatien, 97
visualization, 95–97, *98*, 99, 106–9; *see also* ADCOP
VR (virtual reality) *see* simulation
VVS (Russian Air Force), 21

Wall, Brandon, 133–34
Washington, Gen. George, 97
weapon design, 191–92
wicked problems, 163–64, 166
WiFi, 144
Wild Bunch, The (film), 91
Wohlstetter, Roberta:
  Pearl Harbor: Warning and Decision, 44–45
World War I, 18, 19
World War II, 1–6, 19, 33, 60
  and atomic bomb, 64
  and intelligence, 44–45
  and Maginot Line, 164, 167
  and Malaya, 111–12, *113*
  and Taiwan, 153–54
Wormuth, Christine, 8

Xi Jinping, 35

Yamamoto, Adm Isoroku, 1–4, 6
Yamashita, Gen. Tomoyuki, 111

Zelenopillia, 14–18
Zelensky, Volodymyr, 43, 131
Zephyr, 148–49